Transactions of the American
Philosophical Society
Volume 85, Part 3

TRANSACTIONS

of the

American Philosophical Society

Held at Philadelphia for Promoting Useful Knowledge

VOLUME 85, Pt. 3

Murchison in Moray
A Geologist on Home Ground

with the Correspondence of Roderick Impey Murchison
and the
Rev. Dr. George Gordon of Birnie

Michael Collie
and
John Diemer

THE AMERICAN PHILOSOPHICAL SOCIETY

Independence Square, Philadelphia

Library of Congress Cataloging in Publication Data

Collie, Michael and Diemer, John
  Murchison in Moray: A Geologist on Home Ground

Bibliography, index, illustrations, maps

  1. Murchison, Roderick Impey   2. Geology, Scotland
  3. Scotland, 19th century

ISBN: 0-87169-853-6   94-78514

For Kate

CONTENTS

LIST OF ILLUSTRATIONS

LIST OF MAPS

ACKNOWLEDGMENTS

We are most grateful for the assistance we have received from Dr. Mahala Andrews, the Royal Scottish Museum; Jean Archibald and Mrs. Jo Curry, Special Collections, University of Edinburgh Library; Susan Bennett; Sara Cole, Royal Scottish Museum; Mr. & Mrs. Cowie, formerly of the Manse at Birnie; John Davie, the National Trust of Scotland; Robert H. Dott, University of Wisconsin; Mr. Euan Gordon, Edinburgh; Jenny Hurst, Geological Survey, Keyworth; Jim Ingles, Elgin Museum; Dr. David Iredale, Moray District Record Office; Ian Keillar; Colin McLaren, Archivist & Keeper of Manuscripts, the University of Aberdeen; Ian Morrison, the Falconer Museum, Forres; Mrs. Christine Sangster, the Falconer Museum; Mr. Smart, Keeper of Manuscripts, St. Andrews University; Dr. John Smith, Department of Geography, University of Aberdeen; John Thackray, Sheila Meredith, Wendy Cawthorne, Marie Powrie, Meriel Spaulding and Elaine Bimpson, the Geological Society of London; Dr. N.R. Thorpe, Glasgow University Library; Dr. Charles Waterston, Edinburgh; Mary Robinette and Gary Addington, Department of Geography and Earth Sciences, University of North Carolina at Charlotte. We are most grateful for having had free access to the Gordon correspondence deposited in the Elgin Museum. Special thanks go to Frank James of the Royal Institution Centre for the History of Science and Technology and to Michael Benton of the University of Bristol for reading an earlier version of this manuscript.

In a survey of more than a hundred of the principal research libraries in the United States we have been greatly helped by archivists, librarians and curators of manuscripts, whose courteous and knowledgeable assistance it is now a pleasure to acknowledge. Research has been funded by the American Philosophical Society and the Faculty Development Fund of the University of North Carolina at Charlotte.

PREFACE

This study of Roderick Impey Murchison's geological work in the north-east of Scotland is a contribution to the history of geology in Britain during the nineteenth century and, in particular, during the period between Murchison's first field trip there in 1826 and the commencement of the geological survey of the region in the mid-eighteen seventies, under the direction of Archibald Geikie. It is designed to complement the important publications of other scholars in recent years, notably those of John Thackray, Martin Rudwick, James Secord, Robert Stafford and David Oldroyd, though the focus of attention here is different from theirs.

Our book is exclusively about the Moray Firth, an area of now classic significance for geologists and palaeontologists that was not treated either in Stafford's *Scientist of Empire* or in Oldroyd's *The Highlands Controversy*. Maps 1, 2, 3 and 4 show the boundaries within which we confine ourselves. There are several reasons for the sharp focus on what may at first seem too small an area of land and water to justify the writing of a whole book about it. The first is that the region presented nineteenth-century surveyors and stratigraphers with a set of severe geologic problems which retrospectively can be seen to have intrinsic historical importance, to the extent that the efforts to solve them significantly advanced the study of geology in Britain. Most geologists of standing visited Moray during the eighteen fifties and sixties. We will show what drew them to a remote region that had not previously attracted the attention of scientists.

The second reason is Murchison himself (Figure 1). Even though, like most of us, he made serious mistakes during the course of a long career, we are not disposed to detract from his obvious achievements in the field of stratigraphical survey. We find him both a fascinating and colorful character, and a remarkable energetic scientific pioneer.[1] His work in Moray shows *how* he conducted his purely scientific research over a long period of time, suggesting

---

[1] The term scientist may have first been used in 1840 by Whewell in *The Philosophy of the Inductive Sciences*: "We need very much a name to describe a cultivator of science in general. I should incline to call him a scientist," and in *Blackwood's Magazine*: "Leonardo was mentally a seeker after truth—a scientist; Corravaggio was an asserter of truth—an artist."

the possibility of a different interpretation of this key figure in Victorian science than the current one.

The third reason for the sharp focus on the Moray Firth has to do with the discovery during Murchison's lifetime, indeed during the period of his active research, of the fossil remains in the Elgin district of extinct reptiles. These finds presented the imagination with a sharp challenge. Had there really been reptiles in the north-east of Scotland during some earlier epoch? If so, what were they like, in what conditions had they prospered, and what had rendered them extinct? These questions were taxing enough, but there were others. Had Murchison's first descriptions of rock strata been adequate? Or would they have to be modified to accommodate the findings of palaeontologists? Had these reptiles really been found in the Old Red Sandstone? If not, in what type of rock had they been found? For a few decades, geology and palaeontology were on a collision course. What happened in Moray hinted at a reconciliation.

In short, the region is of great geological interest. Its complexities were bound to puzzle the first geologists who attempted to study it. And the key figure, Murchison, the pioneer Victorian scientist who was indeed puzzled, indeed severely tested, by the complexities of the region, was himself a fascinating figure whom we believe has not yet been adequately described.

The book is based on three sets of documents: Murchison's letters, field notebooks and publications. It is not necessary to claim that these documents are an *absolutely* reliable guide to Murchison's career, or to the way he thought, for them to be of paramount historical importance, since, in a study such as this one, it seems reasonable to begin by paying close attention to what was actually written or produced. Doing so allows one to see the development of his thought and the processes by which he arrived at research conclusions. We resist the temptation to simplify in order to generalize, preferring instead to approach as nearly as possible to the written evidence available to us, to Murchison's own words, and to his periodic correction of what he had earlier written. In Part One, which is in effect a monograph on Murchison as a field geologist in the northeast of Scotland, we emphasize the ideas of development and process as we describe the various questions that concerned him, and relate those concerns to his correspondence with George Gordon (Figure 2).

In Part Two we reproduce verbatim and in its entirety Murchison's correspondence with the Rev. Dr. George Gordon of Birnie, by Elgin, a correspondence not much used yet by scholars and, generally speaking, unknown. This exchange of letters, whose provenance is described at the beginning of Part Two, is only a small *tranche* cut from Gordon's massive correspondence with many of the leading scientists of the day, including such luminaries as Hooker, Darwin, Lubbock and Huxley, and also contemporary Scottish geologists like James Nicol. Here we are only concerned with Murchison and Gordon; the archive as a whole will be described in the forthcoming annotated catalogue being prepared by Michael Collie and Susan Bennett. Murchison's correspondent, the energetic George Gordon, will be described more handsomely in Part One. He was a dedicated, thoughtful, well-intentioned, well-informed local naturalist who, from countless expeditions on foot, by pony and trap, by train and by small boat, had come to know the shores of the Moray Firth like the back of his own hand. Questions about the region, from whatever quarter, tended to be addressed to him. Murchison knew him for almost forty years, the first known communication from Gordon having been written in 1832 and their first known meeting having occurred in 1840. The surviving correspondence, though incomplete, thus records, charts and celebrates an important working relationship, albeit one that functioned mostly by letter, rather in the same way as modern research teams communicate by telephone, fax and e-mail. One of the factors that allowed this relationship to prosper was their seeing eye to eye on many of the contentious issues that will be discussed in Part One. They seem never to have disagreed about religion, the creation of the world, or evolution.

Although Murchison very much liked the field-work on which his profession depended, indeed reveled in it even during his later years, he was also the metropolitan scientist *par excellence*, perfectly at home in the London world of clubs, societies and associations, and in a variety of ways very much at the center of Victorian professional life. Gordon, on the other hand, was the man on the spot. He read the publications of the Geological Society of London, as well as other journals, but his strength lay in his love of Moray, and his knowledge of it. We present in Part Two all the letters from Murchison to Gordon, and from Gordon to Murchison, together with certain other letters that help to round out the story and sustain the narrative from the time of Murchison's first visits to the shores of the Moray Firth to his death in 1871. We believe that not much is to be gained from the editorial option which we have rejected; that is, of using excerpts from these previously unpublished letters to weave a narrative that represents our interpretation of them. However judicious that interpretation might be, the reader would still be held away from the primary evidence. We are in a position to demonstrate that Murchison has sometimes been misrepresented by selective quotation, and we would prefer not to perpetrate that error ourselves. In our eyes it is essential that the correspondence should be printed verbatim and in its entirety.

Because we regard the letter as a nineteenth century research instrument, not as a literary artefact, and because we wish to reproduce here the entirety of what we regard as an interesting correspondence, our search for other autograph letters has been thorough. We have visited the libraries of the Geological Society of London, the Geological Survey of Great Britain, the Royal Geographical Society, Cambridge University, Dundee University, St. Andrews University, Edinburgh University, the Scottish Record Office, the Royal Museum of Scotland, the British Library, Bristol University, the Royal Institution, the Natural History Museum in London, Haslemere Museum, the Royal Society of Arts, and Imperial College. In 1989–90 we conducted a survey by mail of most major American research libraries and, since then, have worked in the library of the American Philosophical Society in Philadelphia, where we received much assistance. It would be a bold person indeed who said he knew the whereabouts of *all* the unpublished autograph material relating to a given subject, but in the case of Murchison we have done our best to ferret out everything that relates to his work in Moray. Nor was this a mere exercise. It will be seen that Murchison was not a person who, having adopted a certain position, obstinately defended it through thick and thin, but rather that he engaged in a continuous process of enquiry that is fairly represented by his correspondence with George Gordon. He was not always right. He could be obstinate. But his mind was never closed to the possibility of his being wrong.

The second set of documents that have a bearing on Murchison's geological work in the north-east of Scotland are his field notebooks, now—most of them—in the Library of the Geological Society of London. Those that have direct importance for the present study are M/N33a (Murchison's expedition with his wife from Inverness to Aberdeen); M/N37 (with Adam Sedgwick on Skye); M/N38 (Skye to the north shore of the Moray Firth in 1827, but revised in 1855); M/N129 (Murchison in the Highlands in 1855); M/N134 (the Elgin reptiles, 1855 but annotated in 1868); M/N135 (Elgin in 1859); and M/N152 (which includes, incidentally, his going to his old family home at Tarradale (Figure 3)). It has to be borne steadily in mind that these field notebooks were Murchison's primary record of the work he had done, strictly speaking his *only* record, and that therefore, whether the notes, descriptions, sections,

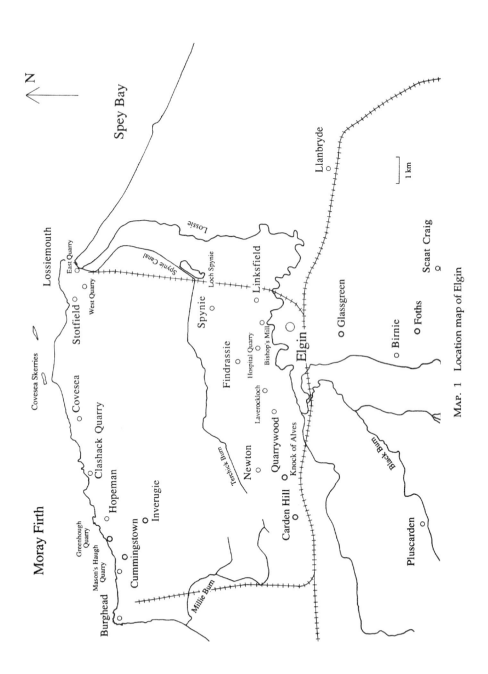

MAP. 1   Location map of Elgin

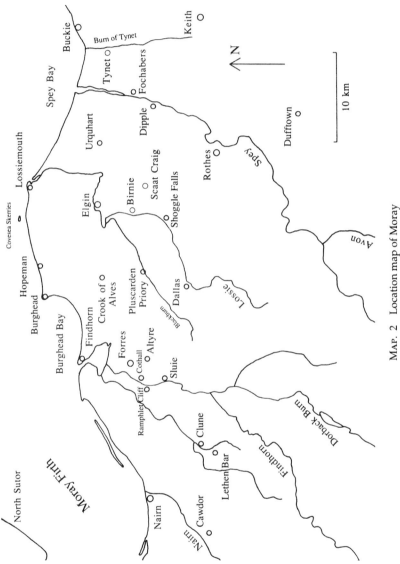

MAP. 2   Location map of Moray

drawings and maps were accurate or not, he *had* to rely upon them until it became necessary to accommodate his own or other people's revisions. They show that the process of revision, reassessment and correction was continuous between his first visit in 1826 to his final use of the notebooks in 1868, three years before his death. They carry the marks of Murchison's attempts over the years to get things right, especially during the course of his revisits to the Moray Firth in 1840, 1855, 1858, 1859 and 1864 to check his data (Map 5). Like most other scientists and scholars, he would naturally have preferred to find, when he revisited Moray and the Black Isle, that his first observations had been accurate. But when he found that they had not been completely accurate, he acknowledged this. If he was pugnacious in defending his opinions, he was also intellectually honest, as the correspondence with Gordon, as well as the notebooks, demonstrate. Murchison called himself "not a palaeontologist but a physical or field geologist" and it is in this light that we shall attempt to give a fair and discriminating portrait of him. In such an enterprise his notebooks are obviously important.

They are more important, we think, than the journal which, in the handwriting of a scribe or amanuensis, is also in the Library of the Geological Society of London. Here we part company with David Oldroyd who in *The Highlands Controversy* gave pride of place to the journal rather than the notebooks when he wanted to follow Murchison's footsteps through the north-west. He based his decision on two things: he considered the notebooks difficult to read and some illustrations had been torn out of them for use in the journal. There seems to be little evidence to support the idea that the actual text of the unpublished journal had Murchison's blessing, whereas, on the other hand, the notebooks show that he gave the scribe, whoever he or she was, considerable freedom. For example, in the important notebook M/N134, which he re-read in 1868, he wrote the word "omit" over the manuscript pages 58–72, apparently leaving it to the scribe to create a continuous journal narrative without reference to them. This example is not an isolated instance. Besides this, we believe it may sometimes be difficult to distinguish between corrections made to the notebooks when early observations were being adjusted and those made while the journal was being prepared. This difficulty will be discussed at greater length in Part One, so that the basis of judgment will be as sound as possible. Meanwhile, for the purposes of this book, we consider that notebook pages omitted from the journal are nonetheless part of the record of what Murchison did when on field trips, which is what we want to know about. David Oldroyd's sub-title is: "Constructing Geological Knowledge through Field-Work in Nineteenth-Century Britain." The journal prepared when Murchison was seventy seven was indeed a construction. The notebooks perhaps provide a better guide to what Murchison did, since they were produced when the actual work was being done. Besides this, because many of the subsequent revisions are dated, they help one trace the development of his thought in a way that the journal does not—though we hasten to add that the journal is by no means overlooked in the present study.

The third set of documents relevant to the present enquiry are Murchison's books and articles, which we list for convenience in Appendix 1. We will refer to and discuss, sometimes at length, all Murchison's articles and notes that relate to the north-east of Scotland. We also attach importance to one of Murchison's principal books, *The Silurian System* or, later, *Siluria. A History of the Oldest Rocks in the British Isles and Other Countries*, published by John Murray and revised three times during the lifetime of the author. In his book *Scientist of Empire, Sir Roderick Murchison, Scientific Exploration and Victorian Imperialism*, Robert Stafford dismisses *The Silurian System* as "a popular text," glossing over the fact that it represented in the first instance seven years of research and writing, and during the next thirty years or so was constantly under review. The three revisions, which it is surely reasonable to consider with some care, represent significant advances in Murchison's thought. Had Stafford's title referred to Murchison's own scientific explorations, he would have had to discuss more extensively Murchison's field-work not only in Scotland, but also in Europe, England, Wales and Russia. But in Stafford's title the term "scientific exploration" refers to other people's colonial explorations, his desire being to demonstrate that Murchison's influence, particularly as exerted through the Royal Geographical Society, had more to do with *gloire*, aggrandisement and Empire than with science. In a book about Murchison's activities in Scotland, there is no need to challenge a sociological thesis of this kind, except to suggest that, in as far as Murchison not only described himself as a field geologist but also remained active in the field throughout the greater part of his life, it seems fair to take seriously the published reports of that work, not least the book called *Siluria*. Whether Murchison was an imperialist manipulator or not, he was certainly an author, so he will be accepted as such in the present book, and his words examined with care. It was actually in the fourth edition of *Siluria*, published in 1867, that Murchison rectified those mistakes to which unsympathetic critics have from time to time drawn attention.

Roderick Murchison was a member of that distinguished echelon of Scots who made names for themselves in London and, indeed, throughout the world. In order to live in London and advance his career, he sold his parents' home at Tarradale just outside Inverness, but he by no means abandoned Scotland,—a matter that will be explored more thoroughly in Part One where an account will be given of the kind of person Murchison was. Advancing his career in London did not entail becoming less of a Scot, something his contemporaries recognized. Chairing a meeting of the Edinburgh Geological Society, Ralph Richardson remarked, as recorded in the *Transactions* of 1874, that Murchison had been "not only a man of science of whose pre-eminence Scotland might justly feel proud" but one who "had always shown the warmest attachment to his native land." Likewise, when Murchison died, the President of the Royal Society of Edinburgh said that "he never to the last relinquished his enthusiastic regard for the land of his birth," sentiments echoed by Archibald Geikie who in the same year expressed on behalf of all the Fellows "the sadness which arises from the recollection of the relation which he bore to the progress of geology in Scotland, and from what he has already done for the advancement of its study in this City." What Murchison had recently done was to donate £6000 towards the founding of a chair of geology at the University of Edinburgh. Earlier he had been awarded the Brisbane Prize by the Royal Society of Edinburgh, the Council minutes for 22 June 1859 specifying that the award was "on account of his contribution to the geology of Scotland." Murchison rarely missed the opportunity to further the careers of his compatriots, whether in Scotland or elsewhere, while as for George Gordon it was Murchison who began the process which resulted in his being awarded an honorary degree by Marischal College in the University of Aberdeen.

As stated, Murchison's work in Scotland began in 1826. In the northeast alone his researches stretched over several decades, though he also knew the rest of Scotland well. In writing the present book, we have therefore come to the conclusion that it is correct to take Murchison's Scottish allegiances seriously and to accept his research interests as genuine. The north of Scotland constituted, for any geologist, a severe challenge—a challenge to which Murchison responded with vigorous enthusiasm, no where more so than in his home territory around the Moray Firth.

Because it seems to us sound practice to incorporate as much of the primary evidence as possible in a book of this kind, and because

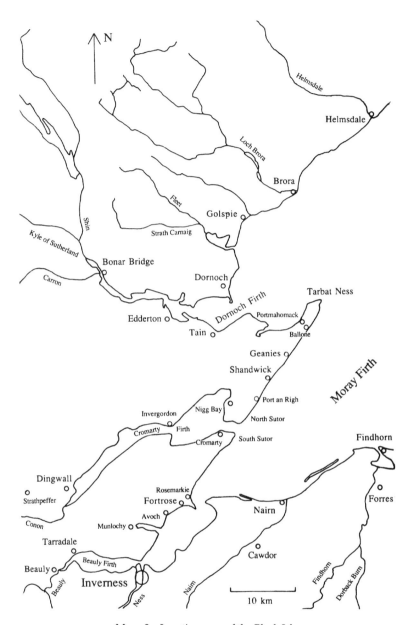

MAP. 3   Location map of the Black Isle

not every one has access to a nearby research library, we conclude
with a set of appendices which we hope will in some measure help
establish the intellectual context within which Murchison worked.
Obviously, in a book about Murchison in Scotland, not very much
will be said about his research in England, Wales, Continental
Europe and Russia. Nor will we examine here his role in the Royal

Geographical Society, important though that was. We are not primarily interested in his public activities, his efforts on behalf of other people, or the metropolitan image he made for himself. It will be enough if we throw a clear light on the way in which he conducted his own work in a place—the Moray Firth—that was important to him, at the same time advancing the study of his subject. The appendix material has been selected to support this aim.

The desire to understand the physical environment in which one finds oneself scarcely requires justification. Not to know what lies beneath the surface of the land, or what the oceans conceal, and what forces that gave the world shape is perhaps a tolerable type of ignorance for some, but for others—those who wish to understand their global habitat—avoiding the problems of geology, geomorphology and palaeontology must seem simply silly or obtuse. Interest in the physical environment does not, in our view, require the support of a theory of knowledge, though the deliberate avoidance of geological problems might. In Murchison's day, the natural desire to think about and understand the physical environment was given additional impetus by a variety of forces, both practical and theological. In practical terms, Victorians would find it useful for construction and mining purposes if the *normal* arrangements of sedimentary strata could be anticipated, identified and measured. Where should railways run? Where would coal be found? In what kind of territory would prospecting for gold be successful? Murchison was frequently asked for his opinion on such matters. Theologically, it is well known that the practical activities, concerns and conclusions of geologists called into question some of the most highly cherished Christian beliefs. Murchison was a practicing Anglican, but he participated in an intellectual process that had an unsettling effect on many of his contemporaries. In other words, for at least these two reasons, nineteenth-century geology is part of our intellectual history. A knowledge of that history is an element in one's awareness of the predicament of the human race.

In a recent correspondence in the *Times Literary Supplement* arising from a review of Alan Gross's *The Rhetoric of Science*, the various participants conducted an argument about the history and sociology of science, some contending that all knowledge is a social construct, others that science "actually captures something of the truth about the world" (John Durant) and is therefore, in some instances, *absolutely* reliable. In the context of this, and similar current debates about the history and sociology of science, we need to say that, in this book, we have little to contribute to arguments about the social production of knowledge. Murchison was undoubtedly interested in establishing the "truth," and would have meant by truth that set of facts to whose existence all thinking, even skeptical persons would have to attest. He would not have been deflected from his desire to get the facts right by the argument that scientific judgment does not depend entirely on evidence, least of all in the northeast of Scotland where debate about the nature of evidence was sustained for several decades.

In the *Times Literary Supplement* correspondence the letter from Christopher Lawrence and Steven Shapin (in the issue of 19 April 1991) contains the following sentence:

(i) neither reason nor the evidence of nature is ever *sufficient* to determine scientific judgments; (ii) what counts as "being rational" and what counts as "evidence" vary from setting to setting; (iii) judgments of validity cannot account for the credibility of scientific claims.

In this new book about Murchison, we do not challenge the utility and intellectual challenge of the discourse analysis that must necessarily underlie any sociological discussion of the production of knowledge, nor, despite the vagueness of words like "determine," "counts" and "setting," do we disagree with the type of statement quoted above. Scientists are obviously not exempt from such analysis; their ideas and formulations do not need to be accepted at face value; they must, indeed, be interpreted; and who knows but that even an interpretation which the reader can see has been organized primarily to confirm the political opinions or class prejudices of the sociologist or social historian may in some way be helpful—to someone. But we do not adopt this approach.

We believe, instead, that Murchison's research activities in the northeast of Scotland concerned real problems, that is, had to do with phenomena that for a while puzzled everyone. We think it interesting to take note, in detail, of what he did, that is, how precisely over a long period of time he conducted his enquiries, visiting the north of Scotland on numerous occasions while keeping himself fully informed about other peoples' research. The continuing *process* of research enquiry seems to us important, so, because Murchison participated in such a process, we will not hold him to ransom for pronouncements made at particular points in time which he later retracted, corrected or modified. On the contrary, we will demonstrate that, though he could be obstinate, outspoken, even imperious, it was nonetheless in his nature to alter his position in the light of new evidence. In this context of continuing research enquiry, we also believe that Murchison's revision of his massive work at first called the *Silurian System*, then *Siluria* is important and worth detailed consideration. Robert Stafford was certainly mistaken when he dismissed this 566 page book as a "popular text," thus absolving himself from the need to take it seriously. Murchison devoted seven years to the writing of it, then revised it three times, these revisions recording the development of his thought up to the fourth edition of 1867, which incorporates, *inter alia*, a new position statement on the geology of northeast Scotland. Generally speaking, we attempt throughout to do justice to Murchison's own texts: his major books, his articles, his field notebooks and, of course, his letters.

PART I

1. THE PLACE AND THE PEOPLE

When the story of Murchison and Gordon's geological research begins some thirty years before the correspondence that makes up Part Two of this book, the Moray basin was a scientific *tabula rasa*. No comprehensive natural history of the region had been written. Its geomorphology had not been investigated. Knowledge was piecemeal and superficial, existing in the day to day experience of local fishermen, merchants, quarrymen and sailors. No one with a scientific training had had a good look at it. Making a living from the land and water was occupation enough for the people who lived there, the principal features of the Moray Firth and the surrounding countryside having remained constant within living memory, and before that also as far as anyone knew.

There had always been castles at Brodie and Cawdor, so it seemed. Cromarty on the north shore of the firth had always been a bustling little seaport. Salmon returned every year to the Spey. Herring were landed at the coastal ports by courageous fishermen, some of whom lost their lives each year in the winter storms. The physical features of the landscape that might alert visitors today to the possibility that the region was geologically interesting had not been noticed, or else were thought about in an old-fashioned way. The essential stability of the landscape as the average person regarded it was endorsed, as it were, by social organizations that also seemed timeless and reliable. The ownership and governance of the land, stretching back in time, and recorded in a mass of legal documentation, was something that people understood and mostly accepted, just as, especially before 1843, they appreciated the stability of the Kirk, taking it for granted, reading or hearing its texts, accepting its observances, and not regarding it quizzically or negatively.

Viewed from Edinburgh, Glasgow or London, Moray was a distant place. The railway had not yet reached it. The Penny Post was a few years in the future. Undisturbed for centuries, the Moray Firth and the Moray basin had not yet attracted the attention of the restless, enquiring, thorough-going scientific intellects that in so many ways characterized nineteenth-century Britain. Certainly many other estuaries and tracts of land awaited the new sort of exploration, but Moray turned out to be special. It had had geological

Fɪɢ. 1 Murchison portrait at Tarradale House

features not found elsewhere in Britain, something impossible to appreciate in the eighteen twenties because no work had yet been done there. Thus when Roderick Impey Murchison began to explore and describe the region, he could not possibly have anticipated the difficulties he would encounter. The North East of

FIG. 2 Gordon portrait

Scotland became one of his field laboratories. It tested his mettle. It put him on trial.

Geological study in the disciplined, professional sense was then in its infancy. When Murchison had his first serious discussions

with Humphry Davy and William Buckland about becoming a geologist,[1] and then continued his self-instruction with the company
of Adam Sedgwick, Charles Lyell, Georges Cuvier and Louis Agassiz,[2] probably only a small number of educated persons could have
identified scientifically the forces that had given the familiar land its
shape. In the years immediately after Waterloo only a few people
had a sense of geological time. In Britain they were constantly reminded of Biblical time and found it sufficient to imagine the progression of progenitors since Adam and Eve, there seeming to be no
reason to challenge or interpret the book of Genesis. For this reason
Charles Lyell, in order to connect distant geological events with
more recent ones, gave his *Principles of Geology* the quaint structure
that it has.[3] The reader had to be taught how to think about geologic
events; how to imagine the deep past.

Because geological information was in short supply (the research
had still to be done), anyone willing to think about geomorphology
had to make a significant imaginative effort, removing himself from
a worldview many of his acquaintances regarded as self-evident.
The old model of the earth had to be replaced by a new one, but the
new one could not yet be described scientifically or in convincing
detail. Geology was a hypothesis in the early stages of being tested.
That it could be tested and that geological truths could be established and demonstrated were the visions that bound together the
first members of the Geological Society of London. They realized

---

[1]In the summer of 1823 Murchison met Sir Humphry Davy when visiting a mutual friend,
Morritt of Rokeby (Geikie, 1875). Davy (1778–1829), Professor of Chemistry at the Royal
Institution and President of the Royal Society, was impressed by Murchison's accounts of
visits to the Alps and the Appenines and he encouraged Murchison to go to London to
pursue science. A year later, in the autumn of 1824, Murchison sold his horses and began
attending chemistry lectures at the Royal Institution. He also attended papers and debates at
the Geological Society in Bedford Street, Covent Garden. When Murchison joined the Geological Society on 7 January 1825, the Reverend William Buckland, reader in mineralogy and
geology at Oxford and canon of Christ Church, was president. The young lawyer, Charles
Lyell, was secretary and had not yet begun his career as an author of geological books. Adam
Sedgwick, professor of geology at Cambridge, Fellow of Trinity College, and prebendary of
Norwich was an active member.

[2]Georges Cuvier (1769–1832) was Professor of Animal Anatomy at the Museum Nationale
d'Histoire Naturelle in Paris and a foreign correspondent of the Geological Society of London. Together with Brogniart, Cuvier made an important contribution to the science of
stratigraphy with their 1811 memoir on the geology of the Paris Basin. Louis Agassiz
(1807–1873) was to make major contributions to the knowledge of geology, palaeontology
and zoology, both in Europe and, after 1846, in America. He began his career as Professor of
Natural History at the Academy of Neuchatel where he was deeply involved in the study of
living and fossil fish. He was also an early proponent for the theory of continental glaciation.

[3]Lyell's *Principles of Geology* was published in three volumes in 1830, 1832 and 1833. It remains one of the most important books published in the earth sciences and it has recently
been reprinted by the University of Chicago (1990) with an introduction by Martin Rudwick
that provides "an outline of the continuous thread of argument that underlies the 1400-odd
pages of the first edition."

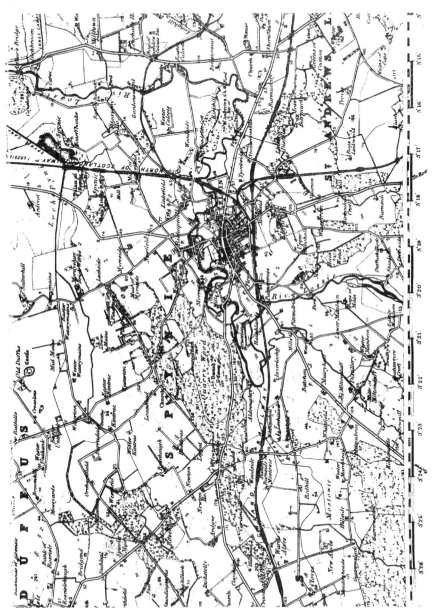

Map. 4 Topographic map of Spynie

that the earth's crust had to be conceived of in a new way and that a lot of detailed work would have to be done before what they imagined to be the case could be demonstrated. It was an exciting time for them. Geology became one of the modern disciplines that any thinking person had to take seriously.

When Murchison joined the Geological Society, it was natural for him to do what the other active members did, that is, identify interesting locations where field work could be conducted in a manner they all would accept. The need for professional acceptance was important. The courteous, but lively cut and thrust of early meetings of the Geological Society gave a person like Murchison the assurance that he needed. If he could satisfy his colleagues within the Society, there was hardly anyone else who had to be satisfied. Murchison first worked in England, Wales and on the Continent of Europe, but being a Scot he understandably decided to conduct some of his research in Scotland, in remote regions for which few topographical let alone geological maps at that time existed, and which only a few of his colleagues knew. The advantage of this was that his research in Northern Scotland was original; he was one of the first in the field. It was also a considerable challenge. The fieldwork took him into wild territory where exploration on foot was arduous and where there were few places to stay. The fieldwork then had to be reported in convincing texts to people who could not immediately verify or even legitimately comment on his results. It is in the relationship of fieldwork to text that we think Murchison's contribution to geology must exist; they are interdependent and cannot conveniently be separated.

We are fully aware, however, that Murchison is a controversial figure, indeed that to some modern scholars he is representative of the worst kind of pre-professional, gentlemanly or dilettante researcher, where the word "gentlemanly" is used in a pejorative sense to mean having access to private money to the extent that one can afford to be careless or unscrupulous in the conduct of one's research. While this judgment cannot be ignored, we have to say the ease with which some university-based critics stigmatize and condemn Murchison, sneering at him whenever the opportunity presents itself, makes us uncomfortable. Murchison became wealthy only after he and his wife had inherited from her mother's estate, more than twenty years after he began his work in Moray. Furthermore we doubt that there is much difference, as far as geological research is concerned, between staying in a castle because you know its owners and staying in a good hotel because you have a generous grant from a research foundation. We have no need to argue the point, however, since our own approach does not involve sociological speculation. Our simple intention is to create the context in which the correspondence that we reproduce in Part Two can best be understood, and we will do this by using his field notebooks and other documents to reconstruct his exploration of Moray, at the same time paying close attention to his revision of the Scottish parts of *Siluria* in its second, third and fourth editions. These revisions record the shifts in his thought based on his own and other people's research over a period of years. *The Silurian System* was first published in 1839 and then, with the new title of *Siluria,* republished in 1854, 1859 and 1867. There is no doubt that Murchison attached great importance to these revised texts because it was not honors, medals and awards but what he wrote that allowed him to get on with colleagues he respected. We will demonstrate the extent to which this was the case. He also attached importance to the papers he read before friends and colleagues at the Geological Society. It is sometimes said that Murchison falsified evidence and presented other people's research as his own. No charge of this kind could be made about his work in Moray. On the contrary, he went there frequently and, while sometimes obstinate in holding to opinions which others challenged, nonetheless he can be seen to have had a sincere desire to check and re-check his results. The reader will be able to judge for himself whether or not this is true because we intend to present the evidence in all its detail. We are anxious, however, that erroneous accounts of Murchison's life should not prejudice the method we have decided to adopt. We will therefore give one example, from the many possible, of how such a distortion might arise—before proceeding to our main task of describing the work in the North East of Scotland that lies behind the correspondence we are reproducing.

The standard and only biography of Murchison is Archibald Geikie's *Life of Sir Roderick I. Murchison Based on his Journals and Letters with Notices of his Scientific Contemporaries and a Sketch of the Rise and Growth of Palaeozoic Geology in Britain,* which was published by John Murray in two volumes in 1875. In a will which Murchison had drawn up in March 1869, he had appointed Tronham Rooks and John Murray to be his literary executors but, presumably after discussion with John Murray, this arrangement was revoked by a codicil dated 10 March 1869, directing the executors "to hand over all letters books and other ornaments necessary for the preparation of the Memoir of my life to my friend the said Archibald Geikie to be entirely under his control." The codicil stipulated that whatever material was borrowed as a result of this instruction was to be

returned to the executors when the biography was published.[4]
Geikie had known Murchison for more than ten years, had accompanied him on a number of expeditions, had collaborated with him
on the production of a geological map of Scotland, and was to succeed him as Director General of the Geological Survey, after being
responsible for the Survey in Scotland. At the time, therefore, he
was the right person for the job and the *Life* he published in 1875 established itself as the official biography. Geikie naturally considered it his duty to present Murchison in a favorable light,
emphasizing his strengths and achievements while playing down
his weaknesses and mistakes. What was wanted was a monument
that commemorated a successful public career and Geikie was in
the position, and had the ability, to provide it.

In *The Highlands Controversy* (1990), David Oldroyd said that

Geikie went far beyond what Murchison might reasonably have expected,
producing a fine account of Murchison's life and times in a beautifully
written biography that tells us a great deal about the history of geological
research in Britain in the nineteenth century. The book was a splendid testimony to Murchison, to Geikie's literary abilities, and to his regard for and
devotion to his former director.[5]

It is impossible to endorse the whole of this assessment which, appearing fairly early in Oldroyd's book, is designed to establish that
it contains all we need to know about Murchison's life. Certainly, it
is a biography which could not be pushed aside with impunity,
since Geikie knew Murchison personally and had in front of him
documents that cannot now be located. Geikie's regard for his former director is impressive. The royalties went to Geikie, not the es-

---

[4]Murchison's will dated March 1869 to his executors Trenham Rooks and John Murray indicates that £1000 be used "for the purpose of arranging as soon as may be all my manuscripts
and journals with a view to the publication of such parts of them as may either illustrate my
own Biography or advance Science but without any liability to account for the said legacies
or the application thereof and I bequeath the copyright of the said manuscripts and journals
and the profits of the publication thereof to the said Archibald Geikie, Rupert Jones, [ ],
William Bates and [Tronham Rooks] equally . . . and on the express condition that none shall
be published without the consent of my executor."

The codicil to the will is dated 10 March 1869 and it revokes the above as follows: "I direct
my executor to hand over all letters, books and other documents in my possession necessary
for the preparation of the Memoir of my life to my friend the said Archibald Geikie to be entirely under his control. He will select from them such portions as may seem to him proper
for publication either in my own words or otherwise as he may see fit and the responsibility
of the preparation of the Memoir of my Life is to rest with him provided always that he or
his representative shall be bound to return to my executor all the letters, books and other documents which they may have delivered to him upon the completion of the publication of the
memoir of my life. . . . "

[5]Oldroyd, David R. 1990. p. 50.

tate, but the biography was in every sense a labor of love. On the other hand, it cannot be called "a fine account of Murchison's life"; too much has been omitted, too much has been glossed over. It is not, of course, a research book; Geikie used the extensive material at his disposal, and he did search out some other material.[6] When incorporating letters into the narrative, as he did frequently, he did not distinguish between complete letter-texts and excerpts from them. The book is without references or notes. In short, and with full acknowledgment that Geikie did well what his publisher expected of him, serious questions have to be asked about the reliability of the account he provided, the more so because the perceived view of Murchison has been based on it, the only full biography that has ever existed. It is natural that this should have happened. Oldroyd said on the page of his book referred to above that "it was Murchison's intention that the journal should be published after his death, and he arranged for Archibald Geikie to carry out this task." This may give the impression that the unpublished journal was fully autobiographical, when it is instead an elaborate attempt to transpose the field notebooks into a continuous narrative. Murchison stated his actual intention in the 1869 codicil to his will.

It may seem odd that our subject, Roderick Impey Murchison, should be introduced indirectly by means of an academic assessment of the only biography of him. There is, however, no greater obstacle to the understanding of a person than the established view of him. With expressions like "Oh, everyone knows what *he* was like," thought stops. In *Controversy in Victorian Geology* (1986), James Secord accuses Murchison of "autobiographical myth-making"—that is, remembering past events in a manner advantageous to himself and disadvantageous to others, in particular Adam Sedgwick. Of this alleged distortion of the record he says "as might be expected," implying that we all see Murchison as a person willing to deviate from the truth retrospectively when his social and professional career seemed threatened. Later in the book he contrasts Murchison's supposed lifestyle in Belgrave Square, and what he claims was a concomitant preoccupation with the "latest superficial sensation," with "men of solid merit" laboring away "unseen," though who these unseen men of solid merit were is not divulged. "There can be little doubt," says Secord, "that Murchison's name was becoming a byword for an elitist and authoritarian vision of scientific activity."

---

[6]In the Haslemere Museum there are letters written by Geikie requesting the loan of Murchison's letters from his known correspondents. There are also letters to newspapers requesting information on articles published about Murchison.

He protects himself with the phrase "whatever the truth of these criticisms," because boldly to assert what is waspishly implied throughout, that Murchison acquired his position in the Victorian scientific establishment by money not work, would be more than he could be expected to demonstrate. Actually, the criticism of Murchison's establishment in Belgrave Square was John Ruskin's, a person whose egoism and mean-mindedness were amply supported by a private income derived from the sale of sherry. Perhaps not enough sherry was consumed at a Murchison soirée. Still later in the book, and still comparing Murchison and Sedgwick, Secord says: "Of the two men Murchison is perhaps the more likely to have constructed a retrospective view of things with the intention to mislead" but then protects himself again by adding "although speculation on such matters is not very profitable." If not profitable, why do it?

Possibly Secord used the word "retrospective" because late in life Murchison had someone help him work through his notebooks with a view to publishing the "journal" of his life, but this journal was never published, being superseded by Geikie's biography. Many writers do indeed review and restructure their memories towards the end of their lives, and biographers have to be careful about what the subject says about himself. In Murchison's case, and for the purposes of the present book, it is the field notebooks themselves that are important and they will be referred to frequently as the unique record of what Murchison actually did, or rather of what he actually wrote while work was in progress. Although his field notes were utilized in formal presentations to the Geological Society, Murchison rarely discarded or misplaced them, and it is easy to understand why. Except for his maps[7] he had no other record of whatever he had done. We find him making mistakes and sometimes subsequently correcting them, or being corrected. We find him forgetting place names and sometimes not being able to remember the location of one of his own drawings or cross-sections. We will not, however, find him abandoning his own notes because of someone else's work in the same locality, except, that is, during the crisis that terminates our story. This belief in his own field notes stems from his having no other records. His research in North-East Scotland stretched over a forty year pe-

---

[7]Geologic maps and cross-sections were an important product of Murchison's fieldwork. For much of the time he worked in Scotland, he lacked adequate topographic base maps. It was not until late in his life that topographic mapping commenced in Moray. For instance, the Rothes 1 inch to 1 mile sheet was surveyed in 1865–71 and published in 1876. The Geological Survey used the topographic maps to record their observations during the first survey several years after Murchison's death.

riod (1826–66) but of course there were many interruptions. He was routinely involved in the affairs of the Royal Society. He was President of the Royal Geographical Society for fifteen years. He traveled extensively in Russia. He devoted much time to adminis-trative duties as director of the Geological Survey. As his many late twentieth-century critics never tire of saying, he had a busy so-cial life. But while all these activities were absorbing his energies, he had his notebooks by him at home as the record of actual re-search in the field that he *had* to depend on. When his interpreta-tions were challenged, or when he became doubtful about them himself, he would put the relevant notebooks in his bag and revisit the problematic field site. This was the way he worked. He de-scribed himself as a field geologist and of course it was part and parcel of being a field geologist that he should produce texts in order to disseminate his results; notebooks, reports, articles and books. He never lost sight that only the rocks he had examined could verify the words he had written. We can best understand him by imagining a constant interior dialogue between a man whose instrument on summer expeditions to places distant from London was a hammer and during long Metropolitan winter nights, the pen.

We also have to take stock of Murchison's manner, which some may find irritating, especially the bellicose ironies of his private communications. Perhaps forgetting that they *were* private commu-nications, some readers have been unsettled by the sabre-rattling quality of some of his pronouncements, no doubt quite legitimately preferring something plainer and quieter, or dull and inoffensive. For example, when Murchison remonstrated with his publisher, John Murray, about the intrusions of an in-house editor, he wrote: "I am sorry that Pentland should have ventured to take so *extraor-dinary* a liberty with my writing as to cut the names of Humboldt, Lyell & Owen out of one of my allusions to *Siluria* & substitute the name of Cuvier alone!!!"[8]

To some readers the three exclamation marks may seem too stri-dent, especially if they suspect the publisher to have had justice on his side. This was Murchison's manner, however, as when many years later he wrote to Herschel:

I am perfectly bewildered in my old age by the new dashing theory of Mr. Croll *"that the earth has passed through several glacial epochs."* He has the audacity to say that *"there are few geologists who will not admit that we have*

---

[8]John Murray archives: 15 October 1850.

evidence of the existence of ice action during the Silurian, Old Red Sandstone &
Crustacean Periods!!!"[9]

Here again the judicious reader, looking at these noisy sentences
out of context, may well conclude, not that Murchison was trying to
put down Croll, but that the number of exclamation marks tended
to increase in proportion to Murchison's fear of being mistaken.
Anyone who wanted to accuse Murchison of having an inflated ego
would certainly latch onto the phrase "has the audacity to say,"
which has the tone of a colonel reprimanding an impertinent sub-
altern, but since *we* have no desire to make such judgments, the accuracy of research results not being a function of self-esteem, we
prefer to direct our attention to the problem of decoding Murchison's manner in order to understand him in his own terms. If this
manner owes quite a lot to the Club and the Officers' Mess, it may
also have a domestic component since irony was one of Charlotte
Murchison's accomplishments also.

Indeed the person with whom he lived might provide a better
guide to Murchison's hopes, aspirations, trials and intentions than
do the retrospective myth-makers. And what, in Murchison's case,
she would have known would have been the energy with which for
years, season after season, he conducted his field-work, and the se-
riousness with which he kept up with his contemporaries, as he had
to, spending hour after hour, each winter, in the study. Rocks in the
summer. Books in the winter. Parties during the London season.
Meetings, commissions and public duties. These are the things
Lady Murchison would have known about in her husband. In the
ordinary domestic context of work and pleasure, one would be ex-
cused for wanting to know what the Murchisons talked about *after*
their guests had departed. Were they infatuated by London social
life? Or were they ironic about it? Meanwhile it seems best to accept
their mode of discourse, colored as it was by a habitual irony that
speaks both of self-knowledge and consciousness of class superiority. When Charlotte says she cannot yet return a book because "her
good for nothing husband" has taken it into the country, she does
not mean this literally. When Murchison calls a new edition of
Siluria his "fat bantling," he did not mean that it was second-rate
or illegitimate. When he says that he is bewildered in his old age,
he does not mean that he is bewildered.

Here, Charlotte Murchison tells a friend, Lady Morgan, about one
of Roderick's expeditions to the North-East. "He has been well, &
very, very gay since his return from the wild shores of Shetland &

---

[9]Murchison to Herschel; 19 August 1867: Royal Society of London HS.12.436.

Orkney," in a letter which incidentally tells us that having passed through Elgin he stayed for a while at Gordon Castle.

At all these places he was a *flirting* cavalier with the pretty ladies, taking them to his dirty stone quarries, & hammering out fossil fish. At the last place they made the discovery of a *certain* monster called by the very sweet sounding name of "Holoptychius Nobilissimus" a yard long!!! What a wonderful history in the life of the said Ladies!!!! & much they laughed at it no doubt. Like poor Lady Cumming Gordon who was to say "I am so *envied* for we live upon the beds of fossil fish."[10]

Here the irony, the word "gay," the exclamation marks, the holding of a subject at a rational distance, and the laughter, may together constitute a class patois to which some might object, but they also place Murchison's geological excursions in the context of normal human endeavor. Thirty years before this letter was written, Charlotte Murchison had herself accompanied her husband to North-East Scotland, so she knew of what she spoke, and thirty years is a long time to accept and support an eccentric husband's preoccupation with fossil fishes, and hammering rocks, and grubbing about in quarries. What she would have understood therefore was that an ex-officer of the Dragoon Guards chipping away at rocks spoke not of a mindless socialite merely diverting himself but of a determined investigator committing himself wholeheartedly to an important occupation. Since she understood, she could afford to be ironic about it. Since Murchison *was* fully committed, there was no need for him to write like a spiritless, grey bureaucrat. He would have his exclamation marks. He would have his style.

Throughout his life Murchison greatly valued professional friendship and the loyalties that go with it. Not everyone liked traveling and working with him, but many people did. Certainly he would sometimes waste a weekend shooting grouse. His wife would quite often tease him about unnecessary indulgences at house parties. Like many other intensely energetic and vigorous persons, he believed he could continue to enjoy claret and port even when his body and his physician were telling him otherwise. Yet these were minor interruptions in a life centered on long, often pedestrian journeys, over land not yet professionally surveyed by geologists. On and during these expeditions he enjoyed, valued and profited from the company of anyone who was equally energetic, and Gordon was certainly one such person, so it is time now to say a few words about him, not least because the correspondence

---

[10]ALS. Charlotte Murchison to Lady Morgan, 22 September [1858]. Beinecke Library, Yale University.

reproduced in Part Two shows Murchison and Gordon being able to work with each other in a friendly way over a long period of time.

George Gordon was born in Urquhart a few miles to the east of Elgin where he was brought up in the comfortable, protected, civilized but not luxurious atmosphere of the Manse. Like many Elgin students he went to the University of Aberdeen, obtaining his M.A. in 1819 at Marischal College, at that time entirely separate from the older King's College. In 1825 he was admitted to the Elgin Presbytery, this being the first public indication that a clerical career had been marked out for him. A brilliant student, it was assumed that he would follow in his father's footsteps. But before and also after 1825 he attended those lecture courses on scientific subjects that were open to medical students at the University of Edinburgh. That was in 1821–23 and again in 1827–29.

In Edinburgh his remarkable talents were fully recognized. He attended the lectures of Professors Graham and Jameson, in the latter case twice, and his lecture notes for these years have survived.[11] While the surviving notebooks may not be a complete record of the instruction which Gordon received in Edinburgh, they are nonetheless helpful in several ways. He attended Jameson's mineralogy lectures in 1821–22, and his geology lectures in the first part of 1822. The geology notebook is endorsed in Gordon's handwriting "Notes from Professor Jameson's Geological Lectures, Natural History Class, Edin. University, 28 January 1822." Interleaved is a second set of notes dated February 1829 and it is in these additional notes that one finds ideas that may have helped him establish common ground with Murchison at a later date. For example, Gordon noted Jameson's claim that a continuity of rock type could be traced from Caithness to Orkney, to the Shetlands, to Norway—an idea that appealed to Murchison after he had been to Caithness (but before he had been to Orkney). Presumably following Cuvier, Jameson also speculated on there being crocodilian but not reptilian remains in sandstone whose fossils were "principally marine."[12] The final section of his notebook entitled "The Physiognomy of the Earth's Sur-

---

[11]Many of Gordon's surviving notebooks are in the archives of the Elgin Museum. The scientific papers of George Gordon consisting of books, letters and notebooks were deposited by Mrs. Gordon in West Register House in Edinburgh in 1979 and were then transferred to the District Archive in Forres where the whole collection was re-catalogued. Each item therefore carries a catalogue number in the form DDW71/ . . . The collection was subsequently moved from Forres to the Elgin Museum.

[12]Cuvier, G., 1813. *Essay on the Theory of the Earth . . . with Mineralogical Notes, and an Account of Cuvier's Geological Discoveries, by Professor Jameson*, First Edition translated by Robert Kerr, Edinburgh: William Blackwood; also Cuvier, G. 1827. *Essay on the Theory of the Earth, by Baron Cuvier, with Geological Illustrations by Professor Jameson*, Fifth Edition, Edinburgh: William Blackwood; London: T. Cadell

face" is one of the more interesting; evidently the student was being challenged by what the professor had to say. We should also notice that Gordon, when he returned to Moray, retained the terminology, recorded on the verso of the first page,[13] that Jameson used to distinguish different types of lithologies. He may have been surprised later when Murchison found it difficult to make such distinctions. In addition to these geology lectures, in 1822 he also attended ornithology lectures, keeping a substantial notebook whose dated annotations show that he kept it by him in later years. He studied zoology also in 1822–23. Annotations to the zoology notebook probably show that he was in Edinburgh in 1825. Finally, it was in 1828–29 that he attended Graham's lectures on botany and Jameson's on geology a second time, having by then a close, friendly relationship with both professors. Having been hard at it for at least eight years, in the summer of 1829 he treated himself to an indirect return from Edinburgh to Elgin—taking in Ben Nevis, Skye, Mull and Iona and, on the way home, Strathpeffer.

Although the educational system at the University of Edinburgh was being widely criticized at that time and Robert Jameson was singled out as being a particularly boring lecturer, this seems not to have worried Gordon, who simply used his years there to become highly proficient both as botanist and geologist. In botany he not only went on field trips with Graham and Graham's students and colleagues, but also with William Hooker, who had recently published his *Flora Scotica*,[14] and was just beginning a distinguished

---

[13]The terminology used by Gordon shows his familiarity with the prevailing stratigraphic nomenclature (cf. Cuvier, 1827 and Lyell, 1830):

2nd Tertiary limestone
Second sand & sandstones
Gypsum & marl
First tertiary limestone—London clay
First plastic clay

4th limestone—chalk formation
4th sandstone. Iron sands. Greensand
3rd limestone. Lias. Oolite. Jura Limestone
3rd sandstone. Variegate Marl. Red ground. [   ]
2nd limestone. Shell limestone. Smoke grey limestone
2nd sandstone. Variegated sandstone. New red sandstone.
1st limestone. Magnesian limestone
1st sandstone. Coal formation

[14]Hooker, William. 1821. *Flora Scotica; or a Description of Scottish Plants Arranged Both According to the Artificial and Natural Methods*. In Two Parts. London: Archibald Constable and Hurst, Robinson & Co. Gordon would have had a great interest in this book, and would have sympathized with what Hooker said in the Preface: "Such a country, though happily now forming an undivided portion of the Empire, is of itself so naturally separate, and was so long regarded politically so, that there can scarcely be raised a question as to how far it deserves the distinction of having a volume dedicated expressly to the elucidation of its vegetable productions."

career that took him first to a chair of Natural History in the University of Glasgow and then to the position of Director of the Royal Botanic Gardens at Kew. When Hooker moved to Glasgow, Gordon continued to correspond with him and later modeled his *Flora of Moray* on Hooker's methods and publications which by then included his *British Flora*. Hooker's letters show that he regarded Gordon as a fully trained, highly proficient botanist.[15] In geology, Gordon's lecture notes show that he had learned a lot from Jameson whatever the professor's failings may have been. Among other things Jameson was one of the principal interpreters of Cuvier, with whose works Gordon became well acquainted. It is worth noting that in these very years (1827–29) when Gordon was consolidating his academic knowledge of geology at the University of Edinburgh, Roderick Murchison, who lacked equivalent academic training, was beginning his exploration of Moray, in 1826 in the company of his wife, and in 1827 with Adam Sedgwick. Although they may not have met until 1840, these two men were to be deeply involved in the study of the geology of North-East Scotland for the rest of their lives.

The intellectual life of the University of Edinburgh was by no means confined to the lecture room. Although it cannot yet be determined with certainty how Gordon divided his time between Edinburgh and Elgin during the eighteen twenties, it is certain that he spent enough time in Edinburgh to establish himself securely as a member of a small group of talented students that included Darwin, Richard Owen, Hugh Falconer, William Stables, John Grant Malcolmson, Alexander Robertson, Ami Boué and Edward Forbes. From their later correspondence one cannot be sure that Darwin and Gordon knew or remembered that they had been at Edinburgh at roughly the same time. William Stables was a local friend; his father was factor at Cawdor Castle to which job he eventually succeeded him. He and Gordon frequently went on expeditions together. Falconer and Malcolmson, both from Forres, were also lifelong friends. They both became Fellows of the Geological Society and the Royal Society of London and both kept in touch with Gordon. All of these people, as well as others one sees in Gordon's correspondence, were students nominally working towards an M.D. Some of them were awarded this degree (Falconer, Malcolmson, Robertson); others had little intention of practicing as a physician or surgeon (Darwin, Gordon, Stables). They were united by

---

[15]See Hooker's "English Letters 1809–30" at Kew, and Hooker's letters to Gordon: 830/1 5 May 1830 and 832/1 3 January 1832.

their unquenchable interest in modern science, acquiring an education at Edinburgh that they utilized and depended upon for the rest
of their lives, whether in India like Falconer and Malcolmson, or England like Darwin and Owen, or Scotland like Robertson, Stables
and Gordon. As students they knew they were the bright sparks.
Several of them became members of the exclusive Wernerian[16] and
Plinian Societies[17] within the university. Gordon's admission to the
Plinian Society in June 1829, sponsored by W. A. F. Browne, Hugh
Falconer, William Stables and Professor Balfour, cannot be taken as
an indication that he never intended to return to Edinburgh. He was
paying tailor's bills in Edinburgh at least until 1832. He was soon to
become a member of the Botanical Society of Edinburgh. The point
is that, despite a small mystery about Gordon's activities between
1829 and 1831, he remained a member of this Edinburgh set. In later
life he and his friends knew they shared a solid grounding in the
two sciences that chiefly interested them, botany and geology.

    Having been admitted to the Elgin Presbytery in 1825, Gordon
undertook his training in theology while in Edinburgh in 1827–29.
In their *Darwin*, Adrian Desmond and James Moore have described
how Darwin, when he went to Edinburgh to study medicine, soon
realized that he was disinclined to follow in his father's footsteps as
a physician—they say he disliked the sight of blood—and that in
casting around for a career that would allow him to continue his scientific work, he actively considered the conveniences of a country
rectory. No one, however talented, could count on a university career or one in government service, both of which in any case had
perceptible disadvantages. To enter the church on the other hand
would give both the time and the security needed for scientific enquiry. George Gordon followed the route that Darwin, having family money, did not, and in his enthusiasm for everything that related
to Moray, soon found ways to reconcile his pastoral duties and scientific interests. After a decade devoted almost entirely to his education, he was offered the Manse at Birnie two miles from Elgin,
being introduced to his congregation in 1832 by the Earl of Moray.
In the same year, now safely installed in the house that was to be his
home for almost fifty years (Figure 3), he sent a first batch of rock
specimens to the Geological Society in London addressed to Roderick Impey Murchison (Gordon, 1832).

---

[16]The Wernerian Natural History Society of Edinburgh was founded by Robert Jameson,
Professor of Natural History at the University of Edinburgh and the leading Wernerian ge-
ognost in Great Britain.

    [17]The Plinian Society "for the advancement of the Study of Natural History, Antiquities,
and the Physical Sciences in General" was instituted in Edinburgh in 1823.

Fɪɢ. 3 Manse at Beirnie

Behind Gordon's *Flora of Moray*, which was published in Elgin in 1839, lies a manuscript notebook in which for many years Gordon recorded his sightings of plants in northern Scotland.[18] Many of the notebook entries are dated, so one can tell where he did botanical work in any particular year. He had explored the remoter parts of Scotland fairly thoroughly. Then, when he returned to Elgin from Edinburgh in 1829, he began to concentrate on the flora of Moray. From that time on, his scientific work was almost exclusively devoted to the region in which he lived. If this started by his walking the length and breadth of his own parish, it soon took him farther afield, so that within a few years he came to be known as one of the people most familiar with that part of Scotland. Furthermore, as soon as he moved into the manse, he began that correspondence of almost Darwinian proportions which kept him in touch with many of the most eminent scientists of the day, particularly geologists and natural historians. Some 1,500 items of correspondence have survived in Elgin alone—evidently only a small proportion of the whole. And it needs to be emphasized that Gordon was no autograph hunter. His correspondence with people like John Lubbock, T. H. Huxley and William Hooker was almost entirely scientific. Very soon, and certainly by 1836, he had helped found the Elgin and Morayshire Literary and Scientific Association, or rather transform it from what it had previously been, and that Society, now the Moray Society, in turn built the Elgin Museum, which was opened in 1843 (see Figure 4). The Museum became the place where people interested in science could congregate. It still stands and contains, not only the correspondence referred to above (letters from scientists and the drafts of Gordon's replies) but also many of the specimens and artifacts which he and his friends collected. The remarkable Museum, which bears testimony to the vision and determination of Gordon and some of his closest acquaintances, continues to serve the social, educational and scientific purposes for which it was founded. The history of the Moray Society will be treated in another study, the present book being exclusively about the geology of the region. It must be said, however, that the speed with which Gordon began his study of Moray, the Moray Firth and the Black Isle indicates pretty clearly what he had in mind when he agreed to accept the living at Birnie. He had in mind to investigate, study, master, write about and thoroughly know a tract of land that no one else had studied in the modern way that his teachers and peers at Edinburgh had shown

---

[18]Gordon, G. 1829. "List of Plants Gathered Chiefly in the Province of Moray." Natural History Museum (Botany Library): 581.9/(411.95)GOR.

FIG. 4 Elgin Museum

him to be appropriate. He was not a country minister who, having time on his hands, drifted into becoming an amateur naturalist; he was a trained scientist who, having chosen to be a country minister, could meet the scientists of his day, if not on equal terms, then at least as a colleague whose wide knowledge was greatly appreciated. In his dealings with them he would occasionally permit himself the gentle ironies that well-educated, disinterested scholars sometimes allow themselves when dealing with goal-oriented, ambitious, career conscious professional scientists, particularly when they affected a Metropolitan or English superiority. But these ironies were mild ones. His letters show that, if he was from time to time patronized by outsiders, he did not very much care, being secure within the intellectual world he had fashioned for himself.

As he said when he retired:

It is only in the very subordinate walks of life that I have done anything of service—namely in natural history and archaeology. What I have done for the Museum has been altogether a labor of love. I have done it for the pleasure of doing it, as a recreation to me, a healthy employment of the body and mind. Or, if I had any ulterior motive in view, it was to work out and make known the natural history and archaeology of the fine old Province of Moray, in which our lot has been cast, extending from the Beauly to the Spey, and from the nearest watershed of the Grampians to the Moray Firth.[19]

As one gets to know Gordon it seems obvious that a natural modesty concealed a burning intellect.

Gordon and Murchison got on well with each other. Although from a biographical point of view one misses those third-person accounts that might have hinted at what they privately thought about each other (Gordon made a strict distinction between private and scientific correspondence), professionally they worked together in a friendly, productive manner until events in the early eighteen sixties strained the friendship. Until then, these unusually energetic men who shared a passionate interest in field geology, happily collaborated in the massive undertaking of coming to terms with the geologic problems that the Moray basin presented, including (1) the identification of rock strata and rock types in a region difficult to survey accurately because of the large masses of drift and the presence of several types of hard to distinguish sandstone; (2) the interpretation of fossil fishes newly discovered along the south shore of the firth; and (3) the resolution of the controversy that arose when

---

[19]*Elgin Courant and Courier*, 15 December 1882.

the fossil remains of previously unknown, extinct reptiles were discovered around Elgin in the confined region between the Findhorn
and the Spey. These, and related problems, were not immediately
apparent. Gordon and Murchison came across them as they worked
together and, increasingly, with other scientists as well. Gordon
published only two articles on the geology of Moray, though it
was one of great importance. Murchison meanwhile regularly reported their findings to the Geological Society, at the same time re-
vising *Siluria* to take account of new knowledge. There is nothing to
suggest that they ever seriously disagreed about the wider implications of geology. Throughout his long life Gordon seldom addressed directly and openly the question of whether geology and
Christianity could be reconciled; public argument about this would
not have been helpful to him. Murchison, for his part, though he
had told Sedgwick in a letter that Geikie suppressed that he had
never been able to share the clergyman's belief in the divinity of
Christ, was nonetheless a practicing Christian who, like many Victorians, preferred to be seen going to church while keeping his ideas
and doubts to himself. His correspondence with Gordon, however,
shows that it became more and more difficult to maintain an easy
going attitude to the conflict between religion and science.

During the years when Gordon was settling into the Manse at
Birnie there were two developments in Moray of great interest to
him. The first was that, as geology became more fashionable, amateur naturalists began to find fossils in the exposed rock banks of
north flowing rivers and burns, as well as in a few other places.
Identifying, classifying and interpreting these fossil finds soon became the business of the Elgin and Moray Literary and Scientific Society. These fossil fishes will be discussed in Section 3. The second
development was that as the population increased in the towns
more and more stone was needed for new housing developments,
especially in Burghead, Lossiemouth and Elgin. In the new quarries
opened up for this purpose a different type of fossil was discovered
that could not at first be identified. This will be discussed in Section
4. The age of the rock in which these finds were made suddenly became a problem. Many of Gordon's contemporaries in Moray had
taken for granted the continuity of the Old Red Sandstone in North-
East Scotland simply because, as Murchison noted in *Siluria*, it was
impossible not to be aware of its extent. No one had expected to find
terrestrial vertebrate fossils of any kind in rock of this age. Had the
rock been wrongly dated? Or did its fossil content point to geologic
events that had not been encountered previously? Finding even
provisional answers to these questions took several decades.

2. EARLY NINETEENTH-CENTURY STRATIGRAPHY: MURCHISON'S INHERITANCE

To understand Murchison's attitude to geological research, one should look at those who wrote about early nineteenth-century stratigraphy. What did Murchison, a self-made geologist, read and to whom did he speak when he began the career that made him famous.

Stratigraphy is the study of layers of rock, or strata, that have been deposited one on top of the other over various periods of time by a range of sedimentary processes. In general, strata are deposited horizontally although inclined layering is also possible. One of the goals of stratigraphy is to correlate rock units over wide areas. In pursuing this, there is a difference between what can be imagined and what can be observed. Was it best, in any as yet unsurveyed region, to start with the assumption of horizontal normality, or was it best to assume that sedimentary layers would usually have been disturbed in some way? If you inclined to the first of these assumptions, you could afford to be not too concerned about the circumstances of sedimentation, looking only at the long-term results. The more interested you became in observable deviations from the supposed norm, the more you would have to think about the possible causes of those deviations. Murchison's research was undoubtedly governed by the first assumption, that the geologist had first to identify the large scale sequences of rock strata, before thinking about what he assumed would be merely local variations to that regularity of sequence.

The point is that he had, intellectually, to start somewhere. What he could not have done in 1825, what no person could too easily have done at that date, was to make a list of the forces that had shaped the known part of the earth's crust. For that matter, such forces or secondary causes were not needed by anyone who believed God must have arranged things as He wished. Consequently, it was easier to imagine regular sequences of rock strata, horizontal in the first instance, than to imagine that surface irregularity or disturbance in a particular region might provide a clue to a geologic situation that had not yet been observed. These early and wholly reasonable assumptions of Murchison's are well illustrated by

Charlotte Murchison's color drawing of the continuity of strata between London and Paris which in her drawing forms a large shal-
low trough or syncline (see Figure 5). She and her husband must
have looked for an example of the idea they wanted to illustrate. No
reproach or judgment is intended here. The immediate purpose is
simply to establish what would have been the predicament of any
person in 1825 starting to think about geology in a serious way, who
began to contemplate rock strata that he knew existed but had not
yet seen. This person might understand the principles of sedimentation in a rudimentary fashion, but lack research or field experience to move beyond the rudimentary. Murchison could not turn to
professionally trained geologists as he would nowadays. He had to
think things out for himself, or try to do so. It might be said that in
1825 he had no reason to suppose that the principal strata, representing vast periods of time, might not be extensive and uniform.

The word 'horizontal' was used in the previous paragraph in a
merely theoretical manner for the sake of explaining what might
be a logical dilemma for a novice who had yet to embark on field-
work of his own. What was he to expect? What would he be looking for? Of course, Murchison even at the outset did not believe
that rock strata would necessarily be horizontal, though late in life,
as well as in the correspondence with Gordon that is reproduced
here, he would sometimes perversely insist upon examples of hor-
izontality. He did believe that inclined planes of strata would preserve the regularity of sequence that he always anticipated and
sought. From his early reading, and from his first encounters with
other geologists, including William Buckland and William Smith,
he understood that in order to study, survey and analyze non-hor-
izontal, though regular strata, he would have to observe and measure both the angle of dip and the strike direction, where the first
"the angle of dip" refers to the degree of tilt of the inclined strata in
profile, and the second "the strike direction" refers to the compass
bearing of a horizontal line contained within the inclined plane.
How did people think about these matters of strike and dip when
Murchison began his work?

In a recent article, J.G.C.M. Fuller (1992) shows how John Strachey in 1719 introduced into geological discourse the use of "stratigraphical cross-sections, or horizontally extended representations
of strata in profile." According to Fuller, Strachey's sections were
new in several ways:

they were drawings rather than lists of strata, they represented true dips,
and most importantly they demonstrated stratigraphic correlation in the

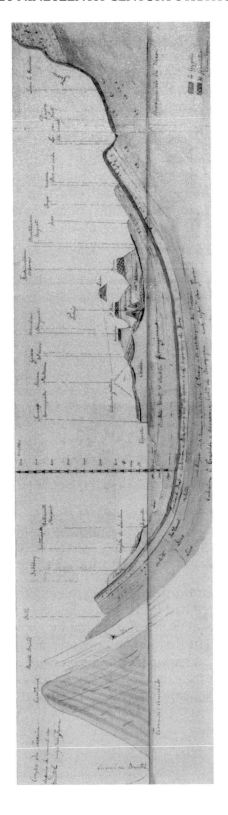

subsurface and across landscape gaps, where for structural or topographic reasons the intervening strata were obscured or missing.[1]

John Strachey's original cross-section as it appeared in the *Philosophical Transactions* is reproduced in Fuller's article and is also reproduced here (see Figure 6). The measurements that Strachey used had been obtained from the published results of bore holes in a coal-bearing region, and they allowed him to calculate the angle of dip—"22 inches in a Fathom", or 17 percent.

Strachey also deduced that between bore holes four miles apart there had been a fault, or what he called a "trap." This explained why the angle of dip at the two drilling points was the same, but the coal strata were offset. This example is used here to demonstrate the sort of thing Murchison would have understood when he began his own work.

In commenting on this cross-section, Fuller lucidly explains the significance of Strachey's preconceptions about regularity:

A really vital and original feature of this cross-section is that its author evidently had no doubts concerning general correlations among strata that were hidden underground, nor of the continuity of individual beds, even across several miles of structural upheaval. Regularity of superposition and lateral continuity of strata seem to have been treated by Strachey as self-evident properties. It was a view which owed more to the colliers' experience, or in his own words, what the "workmen have met with", than with any grand Theory of the Earth.[2]

Another comment of Fuller's has a bearing on Murchison's inheritance from earlier workers in the field. "In Strachey's day," he says,

one must remember the strata had no chronology, only physical presence. That they had been created at the beginning of the world was understood; yet such an understanding called for no sequential or successive creation, no passage of time for the making of one stratum before the next.[3]

A hundred years later Murchison was in roughly the same position as Fuller says Strachey had been; that is, he could feel free to concentrate on the practical utility of stratigraphy for activities such as coal mining, without at first having to worry about the wider implications of geological research. Murchison was not fully a Christian but did believe in the existence of God. As his career developed, he became a successionist, someone who believed that everything

---

[1]Fuller, J.G.C.M. 1992. p. 69.
[2]Ibid., 71.
[3]Ibid., 74.

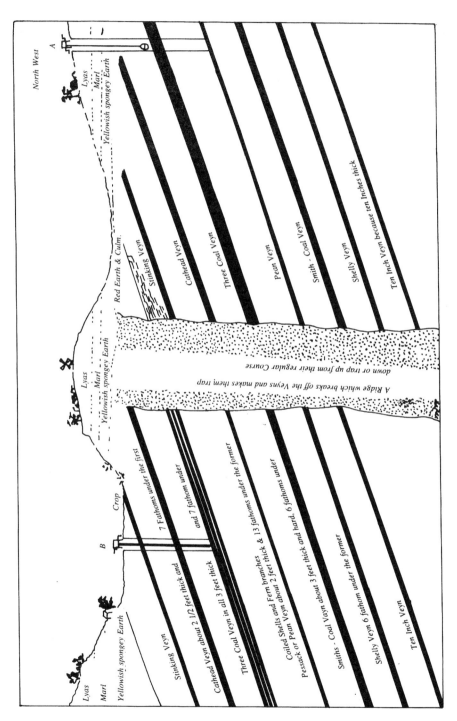

Fig. 6 Strachey cross-section

on earth had been created by God but not at the same time. Only
slowly and maybe reluctantly did he face up to the problems of
chronology to which Fuller refers. Perhaps it would be fair to say
that, in this respect, he was a typical mid-Victorian who found ideas
of stability easier to grasp than ideas of instability.

It will be understood that these references to John Strachey have
been made only for the sake of providing an early example of how
the study of stratigraphy began. It was at first utilitarian in empha-
sis. The threat to religious belief was not perceived. The focus was
upon effects, not causes; on what could be observed and investigated in the present, not on how the surface of the earth's crust had
been formed. Common sense inclined one to a belief in geologic
stability.

When you returned to Britain after thirty or forty years on a different continent, London was where it had been; around your home
town the hills had the same shape; rivers that flowed south before
you left flowed south when you returned; and coal shaft measurements, if accurate to begin with, were still so. Thirty years felt like a
long time. Early Victorian novelists used periods such as thirty or
forty years to allow for significant social change in their fiction, de-
picting superficial environmental or topographical developments
that affected the lives of individuals or families; but few novelists
during Murchison's lifetime gave their characters an awareness of
a deeper or longer world history. Local events seemed not to require
large scale explanations. If a river became clogged with silt so that
ships could no longer use its estuary, no benefit accrued from relating the deposit of silt to any larger geologic overview or theory.

Much of the surface of the globe was being explored for the first
time. A European discovering the Ural or Himalayan mountain
ranges would not, or at any rate mostly did not, ask how they came
to be there. If the history of geology in nineteenth-century Britain
was to involve the progressive abandonment of this sense of geologic stability, matched by a revolution in the way in which
thoughtful, informed people thought about time, it is still worth remembering that when Murchison began his career these developments had yet to occur. For him, stratigraphy was the study of strata
in constant, unchanging relation to each other and his interest was
not in process but in what was there to be observed.

Murchison knew, however, that all this had been boldly challenged by Baron Cuvier.[4] Described in those days as "administra-

---

[4]The relationship of Murchison and Cuvier has been largely neglected in recent books
about Murchison. Stafford does not mention him. Oldroyd mentions him once, in a list.

tor" of the Museum of Natural History, or le Jardin des Plantes, in Paris, and also a professor of natural history at the Sorbonne, Cuvier, by example and by his writings, had revolutionized the study of geology and palaeontology, to the extent that he would have to be taken seriously whether one agreed with him or not. Given the free flow of ideas, publications and people between England, France and Germany between 1815 and 1871, it would be foolish to overlook the international character of scientific work during that period. Though for convenience we will refer to English translations of Cuvier's works, Murchison may well have read them in the original French, since he seems to have been fluent from an early age. Not only do many of his letters show this, he kept his hunting records in French, at least for one season. Murchison came to know Cuvier quite well, exchanged visits with him more than once, and, when he died, delivered an eloquent obituary testimony to his life's work before the Geological Society of London.[5] In the mid-eighteen twenties Cuvier was one of the established authorities to whom you would naturally turn if you wanted advice on any geological problem, which is what Murchison did on a number of occasions, a few of them important. You would be seen to be ignorant if you did not. Consequently, while this is not the place for a thorough-going examination of the relationship of Murchison and Cuvier, a subject that awaits an independent monograph, it will be useful to say a few words about Cuvier's all pervasive influence, because that will allow some of Murchison's ideas about stratigraphy to be identified. If one cannot know exactly what Murchison thought about at this time, one can at least say that the questions that Cuvier raised could not possibly have been ignored. Robert Jameson's annotated 1827 edition of Baron Cuvier's *Theory of the Earth*, was a book with which anyone interested in geology at the beginning of the nineteenth century would be familiar. The third edition appeared in 1817, the fourth in 1822, and the fifth in 1827, exactly at the time of Murchison's initial engagement with geology as a discipline.

Cuvier's *Theory of the Earth* is an expository work, designed in its early chapters to introduce the facts and implications, and therefore the importance, of stratigraphy. He identifies in clear, uncompromising terms many of the basic issues that concerned Murchison for the greater part of his career. We will use his exposition to demon-

Rudwick refers to William Coleman's *Georges Cuvier A Study of the History of Evolution Theory*, Harvard University Press, 1964.

[5] Cuvier's obituary by Murchison in the 1833 Presidential Address to the Geological Society of London as reprinted in the *Proceedings of the Geological Society*, v. 1, pp. 438–464.

strate geological science when Murchison first began to discuss it with people like Humphry Davy and William Buckland. We are not talking here about direct influence, but rather about the prevalence and circulation of sets of ideas at a particular juncture, allowing simultaneously an appreciation of how far people had not gone in their thinking and research. Cuvier acknowledges William Buckland's work[6] and Murchison knew Buckland, who was President of the Geological Society. Murchison did not just set off into the field to look for strata in outcrops, cliffs, quarries and riverbeds. He talked with people first and read widely. Cuvier's *Theory of the Earth* is used here as representative of the ideas that were circulating in Murchison's time, ideas he would have had to come to terms with as he began his work in the eighteen twenties.

In the section entitled "First Proofs of Revolutions on the Surface of the Globe" (where 'revolution' means drastic change or disturbance) Cuvier writes:

The lowest and most level parts of the earth, exhibit nothing, even when penetrated to a very great depth, but horizontal strata composed of substances more or less varied, and containing almost all of them innumerable marine productions. Similar strata, with the same kind of productions, compose the lesser hills to a considerable height. Sometimes the shells are so numerous as to constitute of themselves the entire mass of the rock; they rise to elevations superior to the level of every part of the ocean, and are found in places where no sea could have carried them at the present day, under any circumstances; they are not only enveloped in loose sand, but are often enclosed in the hardest rocks.[7]

By marine productions, Cuvier means shellfish, or rather their fossil remains. How can such remains "the fossil remains of particular species" be found both deep in the earth and high on the hills? If they were all alive at the same time, one would expect them to be found in the same strata. The answer is, he says:

It is the sea which has left them in the places where they are now found. But this sea has remained for a certain period in those places; it has covered them long enough, and with sufficient tranquillity to form those de-

[6]Buckland, William. 1823. *Reliquiae Diluviana; or, Observations on the Organic Remains Contained in Caves, Fissures, and Diluvial Gravel, and on Other Geological Phenomena, Attesting the Action of an Universal Deluge*, London: John Murray. For a discussion of the interaction of Cuvier and Buckland see Page, L.E., 1969. "Diluvialism and Its Critics in Great Britain in the Early Nineteenth Century," in Schneer, C.J. (ed.) *Toward a History of Geology*, Cambridge: The M.I.T. Press.

[7]Cuvier, 1827, p. 6 7.

posits, so regular, so thick, so extensive, and partly also so solid, which contain those remains of aquatic animals. The basin of the sea has therefore undergone one change at least, either in extent, or in situation.[8]

If this had to be accepted as more than a mere opinion, it would be challenging enough. Could one afford to dismiss it? How would someone like Murchison interpret this passage? What parts could he use? Which would he have been unable to use? Horizontal strata, yes; that he could accept. He also accepted that strata, however distant from each other, had probably to be related to each other if their fossil content were the same. He will have a lot to say about this type of fossil evidence in his own subsequent publications. But as for the history of the sea, and the significance of its different levels at different times, he had little to say beyond one of the received opinions of the day that strata must have been raised by volcanic action. Readers of the present book will be able to judge for themselves the extent to which Murchison eventually came to terms with the idea of the sea as an agent of change. Because the book is about his research around an estuary, the question is important.

Where inclined and horizontal strata are found roughly in the same place, Cuvier will only say that all the strata must originally have been deposited by the sea, but that when they were "broken, raised, and overturned in a thousand ways the causes were problematical." Jameson glosses this section in a note of his own on subsidence, where, amazingly, he argues that the "inclined position of the strata is in general their original one" which he claims is consistent with "the great regularity in the direction of strata throughout the globe." Evidently when Jameson wrote his note, the question was considered to be an open one. Experienced though he was, Jameson was not in a position in the eighteen twenties to make global generalizations of the kind he is rash enough to make here, and one does wonder what the idea of regularity of direction means, and how he came by it. We have already noticed, however, in Charlotte Murchison's drawing, a tendency to be more interested in continuity and direction than in what Cuvier calls problematical causes, so one can see, maybe, that they were thinking along the same lines. If strata were not horizontal, some force, *pace* Jameson, must have been exerted to raise one end or lower the other. But what force could that have been?

Cuvier's next introductory section is called "Proofs that such Revolutions have been Numerous," in which he argues that, because all

---

[8]Ibid., 8.

rock beds were originally sedimentary deposits, there must have been not just one but several oceanic events to explain the stratification that resulted. In this passage there is a sentence that may well have attracted Murchison's attention. He says: "The older the strata are, the more uniform is each of them over a great extent."[9] We have already seen that Murchison was more inclined to look for geologic uniformity than what Cuvier calls "variation." Though Cuvier expects uniformity in the oldest rocks, he also appreciates that he has to explain the existence of strata of many different rock types, so he writes what for the purposes of the present book may be an important few sentences:

If we examine with still greater care those remains of organized bodies, we discover, in the midst of even the oldest strata of marine formation, other strata replete with animal or vegetable remains of terrestrial or fresh-water productions; and, amongst the most recent strata, or, in other words, those that are nearest the surface, there are some in which land animals are buried under heaps of marine productions. Thus, the various catastrophes that have disturbed the strata, have not only caused the different parts of our continents to rise by degrees from the bosom of the waves, and diminished the extent of the basin of the ocean, but have also given rise to various shiftings of this basin. It has frequently happened, that lands which have been laid dry, have been again covered by the waters, in consequence either of their being ingulphed in the abyss, or of the sea having merely risen over them. The particular portions also, of the Earth, which the sea abandoned in its last retreat, "those which are now inhabited by man and terrestrial animals," had already been once laid dry, and had then afforded subsistence to quadrupeds, birds, plants, and land productions of all kinds; the sea which left it had, therefore, covered it at a previous period.[10]

Jameson was unhappy with these sentences when he translated them, probably because unable to detach himself from religious controversy about the Flood. Thus he writes a note " Note C, p.334" in which he says: "There are many facts, some of which are recorded in the Bible, which are hostile to Cuvier." Facts?! Although Cuvier had referred not to one but to several marine catastrophes, Jameson felt it necessary to step in to defend the Christian view of the Deluge. We think that this demonstrates once again how questions that might be thought of as resolved or at least satisfactorily reformulated by 1875 were regarded as open and debatable in 1825. Cuvier and Jameson between them pose the question. How would Murchison respond when he began to conduct his research around an estuary?

---

[9] Ibid., 11.
[10] Ibid., 13.

We draw attention, finally, to Cuvier's section called "Proofs that there have been Revolutions Anterior to the Existence of Human Beings," where he attempts to show "that the masses which now constitute our highest mountains, have originally been in a liquid state; and that they have for a long time been covered by waters in which no living beings existed." An orthodox Christian would have found it difficult to swallow this argument, threatening as it was to the anthropocentric beliefs then integral to Christianity. Rock that had once been a fluid was difficult to imagine, even given the example of volcanoes. It may have been difficult, also, to accept the word 'proof'. Some readers may have thought that Cuvier was merely describing Chaos. Whatever the reaction, he takes the edge off this bold, highly imaginative exposition, saying:

Yet, amidst all this confusion, distinguished naturalists have been able to demonstrate, that there still remains a certain order, and that those immense deposits, broken and overturned though they be, observe a regular succession with regard to each other, which is nearly the same in all the great mountain chains.[11]

Discovering the order to which Cuvier refers was to prove difficult, not least in the North East of Scotland.

Whatever Cuvier's reputation at the end of the twentieth century, it is extremely unlikely that many of the early members of the Geological Society of London would have been unfamiliar with his work. Our own argument, however, does not depend upon our being able to show that Murchison was more dependent on Cuvier than, say, Smith.[12] We refer to Cuvier more to note the challenge his works represented as the subject expanded. Impossible, straightaway, for every geologist to take up every aspect of the subject, especially as the means of testing some of Cuvier's assertions, experimentally or by field work, did not then exist. We simply remark that, in a new field in which there was demonstrably so much to be done, Murchison avoided some of the problems others were discussing by devoting his considerable energies to the survey of the oldest rocks as he could then observe them to exist. This itself was a huge enterprise. We will assess later his disregard of causes.

---

[11]Ibid., 18.

[12]William Smith (1769–1839) was an English civil engineer and canal surveyor who used fossils to correlate stratigraphy over much of England. He produced geologic maps and cross-sections using this technique which revolutionized geology and earned him lasting fame. Murchison visited Smith in Yorkshire, worked with him in the field and was indebted to him for early training.

If one wandered about in Sutherland or Caithness in 1825, or tramped around the Black Isle, or along the south shore of the Moray Firth, some great act of imagination would be required to appreciate that familiar topographical features, though stable for centuries, had not always been so. Similarly, if one looked out over the Firth, it would have been difficult to imagine the processes of sedimentation over long periods of time that resulted in those vast mostly invisible strata labeled as periods of tens of millions of years in duration. The behavior of sand would also have been something of a mystery. Several types of sand had been deposited by water long ago. This book in part concerns a controversy about the Old Red Sandstone. This rock could and can be observed throughout the whole region. But the relationship of the other types of sandstone in North East Scotland had not yet been ascertained. Some hypothesis, not just about oceans, but also about fluvial deposits and the action of the wind, would be needed to explain the complex inter-relationship of these various types of sandstone. Interestingly, Cuvier has a section on the "Formation of Downs" which Jameson glosses with Note G "On the Sand Flood." This note introduces a report from a Mr. Ritchie on "Sand Flood in Morayshire." Ritchie takes back his account of the action of wind and sand only to 1097 and he is mainly concerned about how to remedy a local situation by planting trees to prevent further erosion and movement, but it is interesting to notice that Jameson and Ritchie between them were referring to a phenomenon that Murchison was soon to find perplexing. Here again, then, was a situation where one might, but only at peril, attempt to describe what could be observed without speculating about causes. Or should the cause be part of the description? Can there be description without interpretation?

Because of Murchison's familiarity with Georges Cuvier and his publications, we use one of his texts in this section in order to establish the terms of reference relevant to a stratigrapher in 1825. The relationship of fieldwork and text is important. Fieldwork is the *sine qua non* of the geologist. You were not a geologist unless you worked in the field, and Murchison, even during his years as President of the Royal Geographical Society, consistently described himself as a field geologist. At the same time, the only way to become acquainted with what was going on in the field was to read about it. This reading was then put to the test in the field, mostly with varying degrees of success. Like others, Murchison went into the field in the months of good weather. During the winter he read by the fireside and attended meetings of the Geological Society of London. Authoritative textbooks did not exist but, as an alternative to authority, one could meet the people who were tackling the same problems. Murchison made friends easily, taking his professional contacts seriously. In the early days he went on fieldtrips with Adam Sedgwick and Charles Lyell, and later on with Peach, Nicol, Ramsay and Geikie. He never worked in isolation. There was an endless contrapuntal movement of the mind from fieldwork to the text, and from the text to the fieldwork. Gradually, as some aspects of the subject became clearer than others, research interests were more clearly defined, gaps in knowledge noted and discussed. For someone like Murchison, coming to an understanding of geology involved engaging in a communal process of enquiry in which taking a position on any disputed point was secondary to the overall forward movement of research. This does not mean that those concerned did not make mistakes. Everyone did, and Murchison was no exception. Nor does it mean that there were never serious disputes between individuals. There were. It does mean that, in the history of science, identifying what perplexes people is as important as being able to provide, retrospectively, a summary account of a particular person's original contribution to a field of study.

3. FOSSIL FISHES (1826–40)

The first edition of Murchison's *The Silurian System*, containing an important section on North-East Scotland, appeared in 1839. Murchison once said he devoted seven years to the writing of this book, perhaps the winters of 1831–1838.[1] Writing in an age when the publication of a big book was thought to create more of an impression than a string of articles, however much original work they might contain, Murchison went to immense pains over this one. Murchison stopped tinkering with it presumably in 1838. The proofs have not survived, but one's best guess is that he corrected proof during the winter of 1838–39. The dates matter because they incline one to think that Murchison had committed himself to the text of *The Silurian System* before he began to learn of the research being done in the north by Gordon and his friends,' research that will be described in some detail in the second part of this section on fossil fishes. This had the happy consequence of bringing Gordon and Murchison into personal contact in 1840, when Murchison, by then alert that something was stirring in Moray, revisited Elgin. But with regard only to the parts of *The Silurian System* that relate to North-East Scotland (the rest of that great work is not being discussed here), it would be possible to argue that Murchison had committed himself too quickly to an interpretation of the geology of the Moray Firth, and also that he was too dependent on field trips conducted more than a decade earlier, as well as on other people's work on fossil fishes. This dependence was in a sense the penalty he paid for wanting a big book. In working out exactly what happened, we will discount Murchison's reports to the Geological Society, as recorded in the *Proceedings* for 1832 and 1835,[2] and the *Transactions* in 1835,[3] because these were not based on research of his own but on communications received from George Gordon. But we will look

---

[1] ALS. Murchison to James Sowerby, 27 December and 6 March 1838. APS.

[2] As recorded in "Notes and Abstracts from the Minute Book of the Geological Society." On April 1832 was read "Extract from a letter of George Gordon Esq. addressed to Roderick Impey Murchison Esq. FGS noticing the existence of blue clay on the southern side of the Murray Firth." In the Elgin Museum is Gordon's autograph list of rock specimens that Gordon sent to Murchison at the same time. Interestingly, the consignment included three types of sandstone.

[3] *Transactions* Geological Society of London, III (1835): p. 487.

at those articles that resulted from his Scottish visits in 1826 and 1827, because they have a direct bearing on the crisis that is at the heart of this book. Murchison did not visit the North-East of Scotland between 1827 and 1840 (see Map 5). This means that there is a significant gap of time between his original research in the North-East and the publication of *The Silurian System*.

Murchison's parents had lived at Tarradale House just north of Inverness but he had sold their little estate in order to establish himself in London. He would have had a general but perhaps not a close knowledge of the region. In 1826 he and his wife treated themselves to a long, partly social, partly scientific expedition to the North, visiting Edinburgh and Glasgow, and from Glasgow traveling to Arran, Mull and Skye.[4] This part of the journey which lasted for four or five weeks, beginning on 20 June, does not concern us here. From Skye they made the sea crossing to Applecross (home of MacKenzies related to Charlotte Murchison) and from there to Dunrobin Castle by way of Strathpeffer, Dingwall, Bonar Bridge and Golspie. Home of the Duke of Sutherland, this was not the Dunrobin Castle that tourists now visit. The present castle was remodeled at the beginning of the present century when the old one was partly destroyed by fire. Noting that there was nothing in his 1826 field notebook that was not "purely geological," he endorsed it in 1862, thirty six years later, with the comment, "These were the interludes & this was my life and that of my wife."[5] Dunrobin was in 1826 such an interlude; it was their base for field trips northwards, enabling Murchison to do the research that went into one of his more important articles: "On the Coal-field of Brora in Sutherlandshire, and some other Stratified Deposits in the North of Scotland," which was read before the Geological Society on 5 January and 2 February 1827 and then published in the *Transactions*.

It would be difficult to overstate the significance for Murchison's career of this substantial paper, with its table of fossils, its plates and its appendix on the Brora Coal works, followed as it was by a second piece in 1828 entitled: "Supplementary Remarks on the Strata of the Oolitic Series, and the Rocks associated with them, in the Counties of Sutherland and Ross, and in the Hebrides." Having been energetic in the field, he was able to present to his colleagues in London what was undeniably original research, and in later years he would frequently refer to it with the no doubt justifiable pride of someone who believed he had fully come to terms with a difficult

---

[4]See notebook M/N30. The Geological Society of London.
[5]See notebook M/N32. The Geological Society of London.

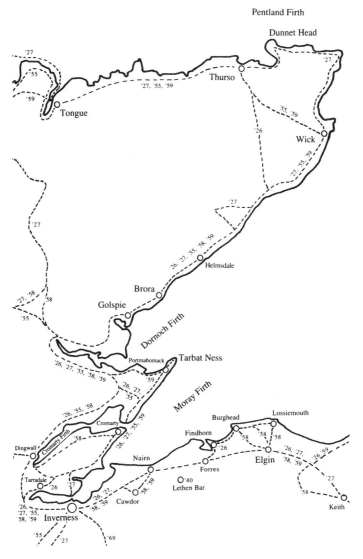

Pentland Firth

Dunnet Head

Thurso

Tongue

Wick

Helmsdale

Brora

Golspie

Dornoch Firth

Portmahomack

Tarbat Ness

Moray Firth

Cromarty

Burghead

Lossiemouth

Findhorn

Dingwall

Nairn

Forres

Elgin

Tarradale

Cawdor

Lethen Bar

Inverness

Keith

MAP 5 Murchison's journeys in Scotland

problem, and in the company of his wife. In the context of the present book, however, the Brora coalfield research was to cause difficulties impossible to anticipate in 1826 and 27. Murchison had looked closely at the Caithness coastline. He knew he had done careful work there. When he and his wife left Dunrobin, they traveled rather briskly through Shandwick, the North and South Sutor, Ethie, Tarradale and so to Inverness. Murchison could plainly see sandstone headlands and strata similar to those he had observed in

Caithness and he therefore assumed a north to south continuity of the Old Red Sandstone along the east coast without at that time doing the detailed work that would have confirmed or modified his overall opinion. In his notebook, he wrote: "In our walk from Balleville to Taradale we found the red sandstone in situ high up on the Muir of Taradale" and a few pages further on "The red sandstone beds dip ENE & decidedly pass under an extensive range of conglomerate hills" and then again "a thick bed of sandstone has been seen at the base of the conglomerate."[6] Murchison did see the sandstone to which he was referring here, but he missed the geological complexity of the land over which he had walked, which was why, as we shall see later, he had to return in 1858.

From Inverness the Murchisons traveled along the coast, most likely using the coast road except when they went inland to places like Cawdor. The present main road did not then exist; in any case it was the coast which Murchison believed would be most instructive.

It is only to the east of the Findhorn that these remarkable landes or sandhills seem to contain nuclei of solid strata & at Burgh Head the strata of the conglomerate sandstone are first seen as the only rocks between that place and Inverness which protrude upon the bank of any part of the Moray Firth—there they dip gently to the sea in a NNE direction.[7]

This and similar notes would seem to indicate that Murchison was looking for what he had already seen in Caithness. He passed through Elgin without an inkling, of course, of how important the town and Moray in general were to become for the geologist only ten or twelve years later. If only he had known what to look for. As it was, the geologist who would say "I stick by my text" had to live with these notebooks as his only research record for the whole period between 1827 and 1858, while, equally, he had no option but to utilize them when he wrote *The Silurian System*. By contrast Gordon's exploration of the same territory was more or less continuous.

In 1827, Murchison retraced his steps this time not with his wife but with Adam Sedgwick, again visiting Arran and Skye, then crossing Scotland to Brora, and from Brora traveling down the coast through Golspie, Shandwick, and Cromarty to Fortrose, then by boat from Fortrose to Inverness. From Inverness they traveled directly to Aberdeen through Forres, Elgin, Fochabers and Keith. The expedition resulted in two joint papers in 1828, one "On the

[6]See notebook M/N152.
[7]See notebook M/N33a.

FIG. 7 Tarradale House

Geological Relations of the Secondary Strata in the Isle of Arran," and the other "On the Structure and Relation of the Deposits Contained between the Primary Rocks and the Oolitic Series of the North of Scotland," this latter being another major enterprise (like the previous year's) consisting of a 44 page printed text, and many plates and figures, including Plate 14 showing the "Murray Firth" (see Figure 8). One can perfectly well understand why Murchison, after this second visit, would believe that no further research was needed in the North-East of Scotland, and we can therefore return to his writing of *The Silurian System*, and in particular to those parts of it that have to do with the North East.

"That fishes existed in strata of the age of the Old Red Sandstone was first pointed out in the year 1827 by Professor Sedgwick and myself," he proclaims at the outset (claiming priority just for Britain, one assumes).

In a geological memoir on the Caithness schists, certain fossil fishes were described which we had obtained from strata, proved by actual sections to form part of the Old Red System of the Highlands, and it was therefore gratifying to find that one of the species which I detected at the very bottom of this system, near Ludlow, was recognized by M. Agassiz to be identical with the *Dipterus macrolepidotus* of the north of Scotland. That celebrated ichthyologist has, indeed, the real merit of having cleared away all the zoological obscurities which previously hung over this branch of our subject. Upon his arrival in this country (to which he was attracted by the well merited honours conferred on him) geologists at once placed all the fruits of their labours in his hands.[8]

Then follows the table reproduced below[9] which demonstrates the extent of Murchison's knowledge in 1839, as well as his considerable indebtedness to Agassiz.

FISHES OF THE OLD RED SANDSTONE

CEPHALASPIS, Agass.
Cephalaspis Lyellii, Agass., Pl.1.f.1
and Pl.2. f.1, 2 and 3.
____ *rostratus*, Agass., Pl. 2. f. 4 and 5.
____ *Lewisii*, Agass,m Pl. 2. f. 6.
____ *Lloydii*, Agass., Pl. 2. f.7, 8 and 9.

CTENACANTHUS, Agass.
Ctenacanthus ornatus, Agass., Pl.2. f.14.

CHEIRACANTHUS, Agass.
Cheiracanthus Murchisoni? Agass. vol. ii. t.1
c.f.3 and 4.
____ *minor*, Agass. vol. ii. t.lc. f. 5.

HOLOPTYCHIUS, Agass.
Holoptychius Nobilissimus, Agass., Pl. 2 bis.

PTYCHACANTHUS, Agass.
Ptychacanthus? (spine of), Pl. 1. f. 9
and 10.

---

[8]Murchison, R.I. 1839, p. 588–9.
[9]Ibid., 589.

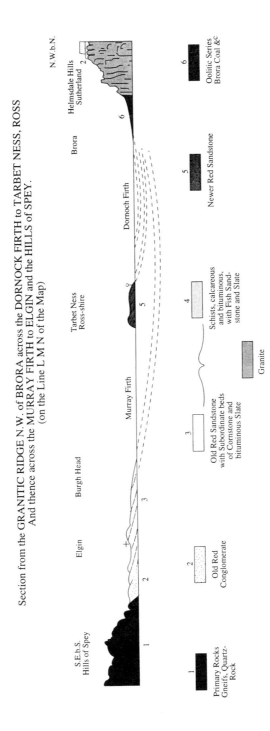

Fig. 8 Geology of the Moray Firth from Sedgwick & Murchison (1828) Pl. 14

ONCHUS, Agass.
Onchus arcuatus, Agass., Pl. 2. f. 10 and 11.
_____ *semistriatus,* Agass., Pl. 2. f. 12 and 13.

DIPTERUS, Cuvier.
Dipterus macrolepidotus, Agass., also Sedgw.
and Murch., Geol. Trans., vol. iii. p. 143.

CHEIROLEPIS, Agass.
Cheirolepis Trailii, Agass. vol. ii. t. 1d and 1e.
f. 4.
_____ *uragus?* Agass. vol. ii. t. 1e. f.1, 2
and 3.

OSTEOLEPIS, Val. and Pent.
Osteolepis macrolepidotus, Val. and Pent.;
Agass. vol. ii. t. 2. f. 1, 2, 3 and 4, and
t. 2c. f. 5 and 6.
_____ *microlepidotus,* Val. and Pent.;
Agass. vol. ii. t. 2c. f. 1, 2, 3 and 4.

DIPLOPTERUS, Agass.
Pro. Brit. Ass. vol. iv. p.75; Agass. vol. ii. p.113.

We will not, of course, hold against him his dependence on Agassiz, except to note that Agassiz was himself feeling his way and in some cases had had only a drawing or drawings in front of him, not the actual specimens. This said, the way in which Murchison writes about his own examples is instructive. The fossil *Dipterus macrolepidotus,* already mentioned, is interesting because Murchison himself found a good specimen while in Caithness with Sedgwick. Here is what Murchison says about it:

Dipterus macrolepidotus, Sedgwick and Murchison.

(Geol. Trans. vol. iii. p. 143. Pl.15. 16 and 17; Agass. vol. ii. p. 117.)

The genus *Dipterus* was established by Cuvier from specimens which I sent to him from Caithness in the year 1827. In the following year, Professor Sedgwick and myself having visited the North of Scotland, we described the general structure of that country, including the black schists and flagstones of Caithness, in which the *Dipteri* are contained, and we then expressed our belief, that these strata were, in part, the equivalents of the Old Red Sandstone of England. That opinion is now confirmed by independent zoological evidence, for I have since detected the above-named species (which is very prevalent in the North of Scotland) in the lower beds of the Old Red Sandstone, at the Tin Mill near Downton Castle, Herefordshire.

This genus was at first separated by Valenciennes and Pentland into 4 species (see Geol. Trans. vol. iii. p. 143.), but after an attentive examination of a great variety of specimens, M. Agassiz, who attempted to class these fishes in the genus *Catopterus,* has definitively concluded, that although the genus *Dipterus* ought to be retained, the supposed four species are only differently modified forms of the same animal.

The generic character of the *Dipterus,* as now confirmed by M. Agassiz, consists in "Two dorsal fins opposite to two similar anal fins, with a caudal fin conforming to that of the genus *Palæoniscus,* in having the vertebral column prolonged into the extremity of the tail."

Having as yet discovered small fragments only of the *Dipterus* in the Old Red Sandstone of the Silurian region, I refer above to the figures of this genus in the Transaction of the Geological Society, and to M. Agassiz's work, vol. ii. tab. 2a. p. 115.[10]

It seems obvious that Murchison is much more interested in establishing, on the basis of fossil evidence, the identity of the Old Red Sandstone in Caithness and Herefordshire, than in doing further work either on the fossils themselves or on the geology of the place where they were found. Congruity was more important to him than difference. This was his Achilles heel.

One of the fishes which Agassiz at that time only knew from a drawing was *Holoptychius nobilissimus*, a fossil destined to play a role in the later part of our story. Perhaps from this example we learn something about the way in which Murchison conducted his research. The specimen in question, measuring two feet four inches by twelve inches, had been found at Clashbennie, near Perth, by the Rev. James Noble (therefore its name). Sir John Robinson showed Murchison a rough drawing of it at the British Association meeting in Bristol. As a result, Noble sent the specimen to Murchison, who had a new drawing made of it. This drawing was used as the basis for Murchison's Figure 1 and was also sent to Agassiz who, without seeing the specimen itself, wrote the following note:

The generic characters consist in the peculiar structure of the scales, the enamelled surface of which is marked by large undulating furrows. Another characteristic feature is in the distant position of the ventral fins, far removed towards the tail, and much nearer the anus and anal fin than in any other genus of the family of *Ganoids*. Lastly, the arrangement of the brachial 'branchiostègues' rays is very remarkable, for they form two large plates between the branches of the inferior jaw, as in the genus *Megalichthys*. They are perfectly well seen in this specimen, and are of a triangular shape. What is perhaps most striking in this Old Red species, is the small size of the head in comparison with the body; for the outline of the two branches of the inferior jaw, which are narrow, are so clearly seen as to enable us to judge of the size or rather length of the head, which was certainly short and obtuse. The structure of the 'nageoirs,' the rounded form of the ventral fin, and the manner in which the rays of its anterior edge are insensibly prolonged, coupled with their relative thinness, are also very marked distinctions, and the same may be said of the anal fin and the disposition of its anterior edge. It is to be hoped we shall at some future day discover a specimen, placed in profile, which will enable us to decide the position of the dorsal fin and the fin of the tail. (The tail is wanting in this

---

[10]Ibid., 599.

specimen.) This individual is lying on its back, and the attitude is very expressive; for it proves that the fish was naturally of a depressed form, 'plutôt déprimé que comprimé,' and not compressed by force. In fact, whenever flat fish are found placed on their backs or bellies in the rock their scales are always deranged, and it is only when they are naturally very round, or even depressed (déprimé), that the ventral scales preserve their natural position as in this example.[11]

There follow some revealing remarks by Murchison himself, evidently inserted into the text at the very last moment as a result of conversations he had with John Grant Malcolmson in April or May 1838. (Malcolmson and Gordon's research in Moray will be examined in detail in the latter part of this section). A late twentieth-century reader will perhaps be surprised by Murchison's willingness to incorporate second-hand information without giving himself the chance to absorb and analyze its implications. He mentions a fossil tooth that Malcolmson said Martin and Gordon had found in the supposed vicinity of the remains of a *Holoptychius nobilissimus*. "Does this tooth belong to this genus or to *Megalichthys*?" asks Murchison. This same tooth, like the scales or scutes of the Holoptychii, turned out to be controversial, but Murchison chose to reproduce it as one of his illustrations, presumably because he wanted to include what other people knew. It would have been wiser for him to have taken more time, especially on a purely palaeontological matter. Then he says lamely: "M. Agassiz must determine the genera and species to which the other remains of fishes found near Elgin may belong".[12] This indeed happened, but Murchison had implicitly committed himself to ideas about the sandstone of the Elgin district even though he had yet to do any close work of his own in that region. This exposed him to criticism which he could easily have avoided. Imprudent to say the least. He himself had walked from Cromarty to Fortrose, and he had been to Ethie; but he had not walked along the rocky, inhospitable shore of the South Sutor from Cromarty itself, or, apparently, in either direction from Ethie. He therefore could not have described in detail what anyone inspecting the cliffs closely would have seen. He might have kept quiet about the "highly inclined strata" until he had seen them; instead he assumes that what Malcolmson referred to must be similar to what he had seen at Gamrie. No wonder that he was back in the North-East only a year after the publication of *The Silurian System*.

---

[11]Ibid., 600.
[12]Ibid., 600–601.

The section on "The Fishes of the Old Red Sandstone" ends with these problematic sentences:

On the other hand, I have recently been informed by Mr. Malcolmson, that Mr. Miller of Cromarty (who has made some highly interesting discoveries near that place), pointed out to him nodules resembling those of Gamrie and containing similar fishes, in highly inclined strata, which are interpolated in, and completely subordinate to the great mass of Old Red Sandstone of Ross and Cromarty.

This important observation will I trust be soon communicated to the Geological Society, for it strengthens the inference of M. Agassiz respecting the epoch during which the *Cheiracanthus* and *Cheirolepis* lived. In the meantime the phenomena at Gamrie may be explained, by supposing that the beds in which the nodules there occur, are regenerated or made up of the detritus of the adjoining Old Red Sandstone. At all events, certain species of *Cheiracanthus* and *Cheirolepis,* as above stated, belong to undisputed strata of the Old Red Sandstone, while no traces of these genera have been perceived in the Carboniferous System.

These two curious genera have not yet been found in the Old Red Sandstone of England.[13]

While Murchison was putting the finishing touches to *The Silurian System* and seeing it through the press, George Gordon was being stimulated to renewed activity in the field of geology by the arrival in Moray of an old Edinburgh friend, John Grant Malcolmson. John's brother, James, lived in London on Wilton Crescent but their mother had remarried and now lived in Forres. Malcolmson had attended many of the same pre-medical lecture courses as Gordon in Edinburgh, after which, in 1823, he had taken the East India Company exams and been sent out to India as an assistant-surgeon. There he had a career of great interest, but also great hardship. While in India, he had continued his geological researches, publishing his findings in the journals of the local branches of the Royal Asiatic Society. Although he was eventually promoted to the rank of surgeon and became Secretary to the Madras Medical Board, he chose to return to Britain to complete the writing of his M.D. thesis on diseases of the liver. The M.D. was awarded in 1839. Before that, and perhaps as soon as he was back in Scotland (not later than in 1837), he renewed his friendship with George Gordon.

Malcolmson had by then published a wide range of papers on a variety of subjects, some geological, some medical, including a cutting indictment of the military authorities in India for their callous

---

[13]Ibid., p. 602.

punishment of soldiers under their command.[14] Once back in the North-East, it was natural that he should address the geological and palaeontological problems that were by then attracting the attention of others. The arrival of a well-educated, experienced expert was timely. As on so many other occasions with other people, but this time in friendship as well, Gordon became his willing guide. They visited all those sites where the complex geology of the district could be observed and also those different sites where fossil fishes had been discovered. That other Edinburgh friend, William Stables, was often with them. Malcolmson, Gordon and Stables had read the early publications of Murchison and Sedgwick; they therefore knew that Murchison had visited some of the sites where fossil fish were now being found without appreciating what lay under his feet. There were other places that Murchison had not visited in 1827, such as Scaat Craig,[15] where important specimens were then being found, notably by John Martin of Anderson's Institution. When Malcolmson's up-to-date knowledge of what was going on in London was combined with the excellent local knowledge of Gordon and Stables, the three friends realized that they had an important research opportunity, both in geology, since the region had never been properly described, and in palaeontology, since many of the fossil fish had not yet been identified.

Leaving Murchison's early articles aside, there was not much else in print for anyone interested in the geology of the region to read beyond a section in James Nicol's 1844 *Guide to the Geology of Scotland* and some fairly detailed notes in Anderson's 1834 *Guide to the Highlands*. In 1835, John Martin had won a gold medal from the Highland and Agricultural Society of Scotland for his "Essay on the Geology of Moray," which was published in Volume V of the Society's *Prize Essays and Transactions*, but the essay pre-dates his own discovery of important fish fossils at Scaat Craig and elsewhere. Consequently it is not too much to say that the work which Malcolmson and Gordon did together represented the first thoroughgoing geological exploration of the region by people well-enough trained and well-enough informed for them to know what they were doing.

---

[14]In India Malcolmson published various letters and notes, and a dozen articles on geological subjects, in the Madras and Bombay publications of the Royal Asiatic Society, as well as two prize essays in 1835 on medical subjects.

[15]Scaat Craig is situated close to the boundary between the Middle and Upper Old Red Sandstone, about 6 kilometers SSE of Elgin. It has been an important fossil locality since 1836 when John Martin, schoolmaster of General Anderson's Free School in Elgin, discovered bones and scales of fossil fish there. Recently redescribed bone material from Scaat Craig might represent some of the oldest land-dwelling, amphibian tetrapods (Ahlberg, 1991).

That Gordon's role was not a passive one is important. He was not a mere guide; he was an active investigator. In the manner of their day, field-trips were interspersed with frequent letters, Gordon's to Malcolmson in February 1838 being of particular importance insofar as he used what he called his "long yarn" to discuss the geology of the sites where they had found fossil fishes. For example, he says of one of these sites near the Glen of Rothes (probably Scaat Craig):

It is pretty evident (if we ascend the burn) where the older beds of conglomerate appear, that the fossilbeds lie immediately above the great conglomerate & the oldest beds of the Old Red Sandstone which in this part of the country rests upon Gneiss. We cannot be so sure of the relation of the cornstone beds or [concretionary] limestone around the burn of Elgin, as between their locality & the first outcropping of the limestone there is upwards of two miles of mere alluvial surface which effectually screens from our view the line of the junction. But were we to hazard an opinion it would be that they are inferior to the limestone.[16]

The length and detail of the letter from which this quotation is taken, together with the roughly drawn cross-sections that are incorporated, sufficiently demonstrate Gordon's independent enthusiasm for the subject. The first field trips of Malcolmson and Gordon took place during February 1838, before which Malcolmson had been by himself, or with Patrick Duff, John Martin or William Stables, to Scaat Craig, Sluie and Linksfield, as well as to Cromarty. Malcolmson soon found that Gordon knew more even than his knowledgeable local friends, as can quickly be seen from the principal record of their field work: the five letters from Malcolmson to Gordon, and the two from Gordon to Malcolmson, all of these being now in the Elgin Museum. There is a close relationship between these letters of 1838 and Malcolmson's first report to the Geological Society of London. Malcolmson had these letters before him, that is, the letters as received from Gordon, not the drafts now in the Elgin Museum—when he wrote the first of his two articles for the Geological Society of London. He also took with him to London specimens collected by Gordon. Whatever he wrote about the geology of Moray on that occasion therefore represented the collaborative research that Gordon most enjoyed.

Unfortunately neither the article itself, nor the illustrative material that accompanied it, was communicated in full to the Geologi-

[16]ALS. Gordon to Malcolmson: DDW 71/838/6.

cal Society. Only after Malcolmson had crossed over to France to consult Agassiz did Murchison present a synopsis of it to the Geological Society. That was on 25 April 1838, when Murchison chose to relate what Gordon and Malcolmson had done in Scotland to what he and others had earlier done in England, thus somewhat disguising from his colleagues, and perhaps blurring in his own mind, the distinctive research challenge represented by the geology of Moray. What Gordon and Malcolmson thought about Murchison's mediation or interference is unknown. The publication of the summary presentation did not of course appear immediately, so Gordon's first hint of what had happened was a brief paragraph in *The Athenaeum* for 5 May 1838 of which he made several copies to distribute to his friends.

It read as follows:

The country around Elgin consists principally of old red sandstone, but at Linksfield, about a mile south of the town, that formation is overlaid by a series of beds, formerly to be considered lias, but which Mr. Malcolmson has ascertained by their organic remains to belong to a freshwater deposit of the age of the Wealden of England. The succession of the strata in descending order is as follows: 1. Blue clay with thin beds of compact shelly limestone. 2. Beds of limestone and clay. 3. Blackish clay. 4. Compact grey limestone with shells. 5. Green clay. 6. Red sandy marl enclosing rolled pebbles or granite, gneiss etc. also angular fragments of old red sandstone. Among the fossils are *Cyclas media,* a common shell in the freshwater strata of Sussex; an *Avicula,* which occurs in the lower purbeck beds at Swanwich, also remains of fishes, and great abundance of a new species of *Cyprus strata*; the equivalent of the Wealden of England, were discovered in the dale of Skye by Mr. Murchison in 1827. The Revd. G. Gordon has recently found the Linksfield fossils at Lhanbryde, three miles to the eastwards of that locality; also a *Pinna,* considered by Mr. James Sowerby to resemble closely a species belonging to the Portland sand; and in making the canal, by which a great part of the Loch of Spynie was drained, fossils were found belonging to the coral rag (?) and the lias of England. Mr. Malcolmson therefore hopes that many of the formations above the old red sandstone, hitherto undetected in that part of the kingdom, will be discovered. He also announced that Mr. Martin, of Elgin, has found in the old red sandstone of that neighbourhood, among other remains of fishes, scales identified with the old red sandstone of Clashbennie. The paper concluded with an account of a series of raised beaches on the adjoining coast, from one of which, fifteen feet above the high water mark, the author procured eleven species of existing *testacea*.[17]

---

[17]Report of LGS meeting in *The Athenaeum*. DDW 71/838/10. This fair copy is not in Gordon's hand.

Although this rather garbled summary is clearly inadequate, and may be unreliable, there are several points of interest. The first is the attempt to establish connections between strata in different parts of Britain. Because Malcolmson is not known to have worked at the English sites, it seems likely that the comparisons were the result of discussion, even instruction, in London, the information he brought south from Moray being interpreted in a manner acceptable to the vice-president. What had been observed in Scotland had to be identified with what was already known in England. Second, the phrase "among other remains of fishes" is strangely casual considering how little was then known in England about fossil fish in the north of Scotland. One might reasonably have expected more excitement or, if not excitement, greater curiosity. But maybe this tells us that Gordon and Malcolmson's real work on fossil fishes was done later, that is, between this first report of Malcolmson's to the Geological Society and his second report a year later. Third, whoever wrote this report for *The Athenaeum* seems not to have known where the places mentioned actually were, a problem that was often to bedevil research in Moray when it was conducted from London. Nonetheless, Gordon and Malcolmson would have known that they lacked the knowledge to interpret the fossils they and their friends were beginning to find. To the extent that their first report to the Geological Society was a failure, or at least a disappointment, they resolved to remedy this by consulting Louis Agassiz, whose *Recherches sur les poissons fossiles* began to appear in 1833 and who in 1834 had been awarded the Wollaston Medal by the Royal Society of London for his pioneering research. Since Agassiz was considered the great authority, he would have been consulted on questions of identification and classification that could not be answered in London. Malcolmson knew that his mentor, Robert Jameson, had already sent his own collection of fish fossils to Agassiz to be identified, described and named,[18] so it was natural for him to do the same. This is why he went to France in April. Murchison later protested that he had not been able to understand what Malcolmson said in a letter from Paris that year,[19] but Gordon would have.

Unfortunately there are no surviving Malcolmson letters to Gordon from Paris, perhaps because he knew he would soon be back in Scotland. When he returned they renewed their exploration of Moray with great enthusiasm, concentrating more on fossil fishes.

---

[18] ALS. Jameson to Buckland, 25 March 1835. Royal Society of London 251/64: "You should see my fossil fishes from the middle district of Scotland which I presume are in the hands of Agassiz."

[19] ALS. Murchison to Agassiz, 27 June 1838.

Although it cannot be demonstrated with complete certainty when Malcolmson was in Forres, when in London and when in Edinburgh completing his thesis, there is no doubt that he and Gordon continued their exploration of Moray through 1838 and during the first months of 1839.

Many years later Gordon recalled the golden days that he and John Grant Malcolmson had spent together:

Well do we recollect the many excursions in which we fondly joined him—such as by the shores of the Firth, tracing its ancient levels, its regular and far-spreading shingly beaches—by the margin of the Loch of Spynie, picking up the remains of its salt and of its fresh-water mollusca, and marvelling at the extent of its now extinct oyster beds—by the rocky channels of the Lossie, and by those of its tributaries, marking in them the contortions of the more ancient rocks, the bluff projections of the conglomerates and their sandstones; here, in their fractures, with angles well defined and sharp as at first, there the rock crumbling at the slightest touch—or by the upper stretches of moorland, finding on them traces of morains and lines of smoothed boulders, resembling what he had seen among the Alps. Never can we forget the entrancing joy with which nodule after nodule was broken, on 14th November 1838, at Dipple, on the Spey, as all the while evidence of embedded organisms slowly, gradually, surely increased, as the light of the rising sun. Great, however, as was the pleasure of discovering fossil fish in this locality, where, indeed, they are in the worst state of preservation, that pleasure was exceeded on a subsequent day when we came across those at Tynat, where the fish are well defined and in a better state of preservation than in any other locality in the north of Scotland. Residing in its immediate neighbourhood, Dr. Malcolmson paid many visits of discovery to the peerless banks of the Findhorn, where the largest sections of the strata are to be met with. Accompanied by Mr. Stables, of Cawdor, he examined the valley of the Nairn, and that of the Burn of Brodie. Perhaps Dr. Malcolmson's noble heart reached the acme of scientific satisfaction on 27th March 1839, when his friend and fellow-labourer, Mr. Stables, laid open in his presence a nodule, at Lethen Bar, that first revealed a form distinct from any previously known fossil.[20]

Malcolmson reported by letter that Stables "got a magnificent *Coccosteus,* showing the tail and wings, and also a fine small specimen of the species with the large tubercles, showing the tails as I saw it in some of Traill's Orkney fish."[21] The excitement mounted. The next day Malcolmson wrote to Stables: "The more I think of our

---

[20]Gordon, George. 1859, p. 28.
[21]ALS. Malcolmson to Gordon. DDW 71 839/3.

discovery the more important it seems. You must let me have the two tailed species of *Coccosteus,* and the large tuberculated fish found by Dunbar, to have drawings of them made; and the first, I fancy, must be sent to Agassiz."[22]

By the beginning of May, Malcolmson, with many of the specimens in his luggage, was in London where he discussed their finds with Murchison who "was astonished beyond measure" and "urged an immediate memoir," one which Malcolmson was only too willing to write. In one of his letters to Gordon, Malcolmson reported that Murchison had said that he and Sedgwick had had "no notion that organic remains lay under their feet."[23] What Malcolmson called "the winged creature," still unidentified, was of particular interest to Murchison and his friends. Because the first volumes of *Recherches* had already appeared, and Agassiz indeed had been to London and was personally known at the Geological Society, it was natural for the Londoners to turn to him. He and Murchison were in correspondence to ensure that *The Silurian System* and *Recherches* were consistent with each other. But it seems to have been Malcolmson who took the fossils to the Continent, together with a box of specimens from Hugh Miller.

During these comings and goings, Malcolmson wrote the paper reporting their two seasons of field-work in Moray—no doubt the most thorough investigation that had been conducted up to that point in time. As soon as he had delivered it, he reported to Gordon that all had gone well:

The fossils I presented in our names, jointly with Mr. Stables, and I propose leaving most of the specimens here for the present, as Agassiz is expected at the end of the summer; and those belonging to Martin, and such of our own as it may be expedient to keep for the north, can be packed up after he has had the use of them.[24]

Fortunately Malcolmson sent a manuscript copy (perhaps an early version) of his paper to Stables. "Fortunately," because the paper was never printed in the *Transactions of the Geological Society*, indeed was lost sight of for almost twenty years, though Gordon never forgot it.

In 1840 Malcolmson returned to India, no longer in the employment of the East India Company, but as partner in the firm of Forbes & Co. of Bombay, of which his brother James was already a

[22] ALS. Malcolmson to Stables. DDW 71 839.
[23] ALS. Malcolmson to Gordon. DDW 71 839/7.
[24] ALS. Malcolmson to Gordon. DDW 71/839/8, 12 June 1839.

partner. As far as is known he never returned to Britain. If he kept in touch with Murchison, Gordon and Stables, as he probably did, the correspondence has not survived. Given the time it took for academic journals to be published and for news to travel from Britain to India, he may not have been unduly bothered by the non-appearance of his paper, especially as he knew the fossils had been put to one side until Agassiz could examine them. In his spare time he renewed his geological exploration of territory within range of Bombay, became editor of the *Journal* of the Bombay Branch of the Royal Asiatic Society, and contributed two long resumés of Murchison's *The Silurian System* and the early volumes of Agassiz's *Recherches sur les poissons fossiles* to the *Calcutta Journal for Natural History*, of which he was a founding member. Presumably he had come to terms with these massive works on his long sea voyage from Marseilles to Egypt and from Egypt to Bombay. Sadly, however, Malcolmson died on a field trip in 1844, without having had the pleasure of seeing the research he had done with Gordon and Stables in print. The manuscript of his paper in fact remained in an office at the Geological Society until in 1859 Gordon prevailed upon Murchison to search for it.

Among those people who are unfailingly critical of Murchison, there have been one or two who have suspected that he knowingly suppressed Malcolmson's findings because they had a direct bearing on his own work around the Moray Firth. Gordon himself briefly harbored such a suspicion, which, however, is not supported by the evidence. On 22 May 1840, stranded off Cuxhaven while one of the steamer's paddles was freed from a rope that had become entangled around it, Murchison wrote to the President of the Geological Society in the following terms:

In the hurry of leaving London it occurs to me that a report I had written out on Dr. Malcolmson's Memoir which is to be laid before the Council at your next meeting was mislaid, & I therefore write these few lines as a substitute.
   The Memoir is highly worthy of a place in our Transactions, because it adds many good data to a subject already much developed a few years ago by Prof. Sedgwick & myself. The chief new feature of the Memoir is that it shews how the middle formation of the Old Red System is tenanted by certain species of ichthyolites, the lower division by others.[25]

Whatever else one might think about this letter it would seem to absolve Murchison from any charge of duplicity but he none-

---

[25]ALS. Murchison to Buckland, 22 May 1840. Geological Society of London, LDGSL 768–18.

theless soon forgot about the whole business becoming, as he explained later by way of justification, very busy indeed with his research in Russia. There was nothing in Malcolmson's work, as best he could recall, that challenged his overall views about the continuity of the Old Red Sandstone. He did not feel threatened by it. He was not a palaeontologist but a field geologist. As we have already seen, his remarks about fossil fish in *The Silurian System* were largely derivative. He did not need additional information from Malcolmson or anyone else to support what he had already said in a manner that satisfied him. Despite the "many good data" referred to above, Murchison chose, when he revised the book for a second edition, to be content with omitting the derivative, out of date, and in parts irrelevant table reproduced above, but did not, as well, include information that had become available since 1839 especially in the publications of Agassiz and Hugh Miller. Gordon would obviously have noticed this.

The partial reconstruction of the field trips and publications has been possible because Gordon kept many of Malcolmson's letters (those referred to above are still in the Museum's archive) and Stables kept a fair copy of Malcolmson's 1839 address to the Geological Society of London. Gordon referred to it twenty years later. During those twenty years, Gordon continued the work he had begun with Malcolmson, sharing his findings with friends like John Martin and Patrick Duff, and in his usual studious way studying with care Agassiz's *Recherches sur les poissons fossiles* so that he could determine which genera and species were represented in Moray. Some of his work-papers on this subject have been preserved in the Museum. Many of the actual specimens they collected are there as well. In his 1839 paper Malcolmson said (according to Gordon):

I have ascertained, by comparisons with numerous specimens, that they (viz. the fossil fish of the Old Red Sandstone of Moray) belong to the same genera and species which are found in Orkney, Caithness, Cromarty and Gamrie—the most common belonging to the genera *Dipterus, Diplopterus, Cheiracanthus, Cheirolepis, Osteolepis*—; a very important fossil, belonging to an undescribed genus of Ray, to be named by M. Agassiz, *Coccosteus*; and the singular creature figured (No.   ), which, notwithstanding its anomalous form, I believe to be a fish. Regarding this very remarkable fossil I propose to submit some observations at another time; but I would now call attention to the part of the bony structure on which the arms rest, resembling that of some fishes of the cornstone series and mountain limestone, and to the resemblance in form of the convex plate on the back of those of the Findhorn and Clashbennie already referred to.[26]

The first part of this statement corresponds with Murchison's table of "Fishes of the Old Red Sandstone" on p. 589 of *The Silurian System*. One supposes it would have been entirely acceptable to him, since Malcolmson was confirming his own findings about the continuity of the Old Red Sandstone from Orkney to Moray. The two lists do not correspond exactly because some of the fish fossils found in other parts of Britain had not yet been detected in Moray, and because some of the fossils that had been found in Moray had yet to be named. Were these fossil fishes of intrinsic scientific importance, or were they only important to Malcolmson and Murchison because they confirmed the continuity of the Old Red Sandstone? No palaeontologist, Murchison regarded fossil fishes chiefly as stratigraphic evidence, so it is possible that Malcolmson's paper put him at his ease more than alerted him to problems that he had not yet thought about. With his customary bonhomie and largesse he supported Darwin in recommending Malcolmson for election to the Royal Society of London.

How Gordon re-constructed Malcolmson's 1839 paper when he recovered a version of it from Stables in 1859 is important. He re-checked parts of it—these will be discussed in the next section because they represented work he did himself. Other parts he quoted verbatim, as here when he reproduces what Malcolmson said in 1839 about fossil fishes in the Stables copy. This is possibly because he saw that the list of genera was less than satisfactory, that is, it was not an accurate indication of the species to be found in Moray. Quite understandably, he was able in 1859 to produce a list of fossil fishes, with locations, that went far beyond Malcolmson's and Murchison's knowledge in 1839. While Murchison was busy with other matters in the south, and overseas, Gordon resolutely continued the study of his native province, step by step making it more complete. It seems likely that when they actually started working together, Murchison did not at first understand the extent of Gordon's knowledge. He had gained it in the field, unlike Murchison, much of whose palaeontology was second-hand.

Gordon's 1859 list follows:

Pterichthys latus, Lethenbar.   *Holoptychius giganteus*, "Dans le couches
Pterichthys Milleri, Clune.   de Old Red en Escosse à Elgin."

---

[26]Gordon, George. 1859, p. 21–22.

Pterichthys productus, Lethenbar.
Pterichthys cornutus, Lethenbar.
Pterichthys major, Findhorn
  and Scaat Craig.
Placothorax paradoxus, Scaat Craig.
Coccosteus oblongus, Lethenbar.
Coccosteus maximus, Lethenbar
  and Boghole.
Acanthodes pusillus, Tynat.
Cheiracanthus microlepidotus,
  Lethenbar.
Diplacanthus straitalus, Lethenbar.
Diplacanthus longispinus, Lethenbar.
Cheirolepis Cummingiœ, Lethenbar.
Osteolepis major, Tynat and
  Lethenbar.
Diplopterus macrocephalus,
  Lethenbar.
Stagonolepis Robertsoni (a reptile?),
  Lossiemouth.

Holoptychius nobilissimus, Elgin.
Actinolepis tuberculatus, Findhorn.
Dendrodus latus, Scaat Craig and
  Findhorn.
Dendrodus strigatus, Scaat Craig.
Lamnodus bifurcatus, Scaat Craig.
Lamnodus Panderi, Scaat Craig.
Lamnodus sulcatus, Elgin.
Cricodus incurvus, Scaat Craig.
Asterolepis Asmusii, "Environs de
  Elgin."
Asterolepis minor, "Environs de
  Elgin."
Asterolepis Malcolmsoni, Elgin.
Bothriolepis favosa, Elgin.
Bothriolepis ornata, Monachtyhill,
  and also near Nairn.
Cosmacanthus Malcolmsoni,
  Scaat Craig.[27]

The interpretation of this list is far from simple. Because of the extreme instability of Victorian naming, at least as far as fossil fishes were concerned, to have the name without the specimen is bound to cause difficulties. One must assume that the list represents every specimen in Moray that Gordon knew about in 1859. He most likely took the names he used from Agassiz and Hugh Miller. In *Monographie des Poissons fossiles du Vieux grès rouge* (1844–45) Agassiz included in Table 31, Figs 13 & 14, pictures of the two *Stagonolepis* fragments he had been sent from Scotland. Gordon knew that this attribution had been questioned, therefore the question mark, but he had not yet read Huxley's *Stagonolepis* article which only appeared at the end of the year. The French which follows *Holoptychius giganteus* also comes from Agassiz, as does the notation which follows the first two *Asterolepis* entries. Quite reasonably Gordon assumed that Agassiz was the authority on whom he could depend at that time, the more so because he was also following Hugh Miller's usage in *The Old Red Sandstone* (1857) and *The Footprints of the Creator* (1861). Because some of Gordon's names were not used by Agassiz, determining how Gordon identified some of the specimens he had in hand is an open question. The list represents the enthusiasm of the local scientist investigating new sites, so that almost inevitably many of the generic names were open to challenge, while many of those for species turned out to be synonyms. Gordon's lo-

---

[27]Gordon, George. 1859, p. 40.

cal knowledge, great as it was, did not allow him to be perfectly sure about the type specimen.

Nevertheless the list does represent an advance in knowledge. Whereas *Cheirolepis cummingiae* had been figured by Agassiz in Table D, Fig. 4 and Table 12, described by Miller in *The Old Red Sandstone* and listed by Murchison, *Pterichthys milleri*, figured by Agassiz, described by Miller and listed by Gordon, was apparently not known to Murchison. A comparison of the lists of fossil fishes published by Murchison and Gordon, when regarded in conjunction with the descriptions of Agassiz and Miller, would seem to show that, if Murchison had followed the extension of knowledge in Scotland and elsewhere between 1839 and 1859, it was not knowledge that he felt he had to utilize in *Siluria*. Certainly it would be a delicate task to determine what exactly Murchison knew or did not know at any given point in time, for example, when he was preparing the third edition of *Siluria* for the press. On the other hand, it does seem fair to wonder why the discovery of new specimens did not arouse his interest, inducing him to revise and up-date the table reproduced earlier in this section, if not for the 1859 edition then at least for the fourth edition of 1867, since the new knowledge might be thought to confirm his opinion about the continuity of the Old Red Sandstone. Of course, there was a snag and perhaps Murchison perceived it. No one could yet say why the fossil fishes listed by Gordon had been found only in certain parts of Moray and Nairn and not in others.

Of the various colleagues and associates who may have noticed Murchison's weak handling in *Siluria* of his own information about extinct Devonian fishes, there was one who most certainly did. In 1861, Huxley published his "Preliminary Essay upon the Systematic Arrangement of the Fishes of the Devonian Epoch." What Huxley did in this publication was challenge the currently used family and generic names of fossil fishes, proposing instead a new classification of six families (sub-orders) that were exclusively Devonian and two that were Devonian but in his opinion not exclusively so. This rearrangement was based both on his own examination of the anatomical structure or organization of specimens available to him and on recent publications not referred to in Murchison's *Siluria*, notably Pander's *Ueber die Saurodipteren, Dentrodonten, Glyptolepiden, and Cheirolepiden des Devonischen Systems* (1860)[28] and Egerton's "Remarks on the Nomenclature of the Devonian Fishes" (1860).[29] He

---

[28]Foster, M. and Lankester, E.R. 1899. *The Scientific Memoirs of Thomas Henry Huxley*, v. 2, p.477.
[29]Ibid., p.477.

had also heard and understood the implications of the Rev. Dr. Anderson's talk "On Dura Den Sandstone" at the 1859 British Association meetings in Aberdeen which will be referred to again in the next section. On the face of it, therefore, Huxley could be seen to be striding ahead of the competition, as though to say: "If you want to use fossil fishes as evidence for anything at all, at least ensure that the classification, and therefore the names of the order, suborders and genera, are based upon sound anatomical information." It has to be said straight away, however, that neither Huxley's classification nor his names stood the test of time. His division of Devonian fishes into six families "Glyptodipterini, Saurodipterini, Polypterini, Phaneropleurini, Acanthodidae, and Cephalaspidae" was not accepted by his contemporaries, and indeed by the end of the century had become meaningless at least if one takes a text such as A. S. Woodward's *Vertebrate Palaeontology* (1898)[30] as representative of the usage that had gained acceptance by then because based upon a far greater amount of information than was available in 1861.

Huxley did not immediately publish the more extended study of Devonian fishes explicitly anticipated in this "preliminary essay," but his tacit acknowledgment that there was still much more to be learned did not prevent him from introducing into the essay a few wholly characteristic and gratuitous touches. For example, of one set of his own physiological observations he says lamely: "These are the unquestionable facts. I leave their bearing on the great problems of zoological theory to be developed by every one for himself." Well, when Huxley adopted this strategy, as he frequently did, he would at least admit that there were important theoretical problems to be worked out, albeit by other people, which is more than some purveyors of unrelated facts ever allow. On the same page he permitted himself an observation one would have to say was not directly relevant to his ostensible subject. "No vertebrate animal higher in the scale than fishes is as yet certainly known to have been found in any rock of Devonian age. In fact, until demonstrative stratigraphical evidence of the

---

[30]In his letter to Gordon dated 16 November 1858, Huxley had already asked a similar question. Whereas Murchison continues to express an interest in the presence of Holoptychius in Moray in the hope of being able to demonstrate a continuity in the Old Red Sandstone, Huxley's purpose was to establish that the fossil remains of *Stagonolepis* had not been found in the Old Red Sandstone where fish scales had been discovered. The two letters obviously derive from a conversation of Huxley and Murchison's. Their enquiries would have developed differently had they gone together to the district that was now of such great interes to them, but they never did.
[31]Since his geological survey of the region with Sedgwick in 1827.

Devonian age of the well-known Elgin beds is obtained, the bearing of the palaeontological evidence against that conclusion is too strong to allow of its being entertained."[32] Apparently this was how scientists were learning to address each other during the period in which academic disciplines were becoming more professional, and yet in all likelihood Huxley must have been exasperated by Murchison's apparent insouciance. Huxley had already twisted the knife by mentioning that some of his specimens had come from Caithness. Now he virtually challenges Murchison to entertain a radical revision of his long-cherished opinions about the Old Red Sandstone. And he was by no means through with him yet.

[32]See Map 1.

4. THE REPTILES (1840–59)

Although Murchison stayed away from Moray between 1840 and 1858, there were events and publications in those intervening years that ensured that Gordon and his friends continued to think about the geology of the region (Figure 9). In 1842 Patrick Duff published a set of well-received newspaper articles as a book entitled *Sketch of the Geology of Moray* with the result that local people tended to turn to him when they discovered anything of geological interest. Duff's work was known outside Moray, included a rudimentary geologic map prepared by John Martin (see Map 6) and was incorporated in a geologic map of Scotland prepared by Nicol in 1844 (Map 7). In 1844 workmen from Lossiemouth brought Duff fossil remains of what later turned out to be a *Stagonolepis*. But was it a fish or a reptile? In 1851 he was brought the partial remains of a *Telerpeton*, a fossil that was to be associated with the Old Red Sandstone for many years and which would be figured in the 1854 edition of *Siluria*. In 1852 Gideon Mantell wrote about this specimen in "Description of the *Telerpeton Elginense*, a Fossil Reptile Recently Discovered in the Old Red Sandstone of Moray (see Figure 10A). Interestingly, when Lyell commented on this paper, Mantell replied: "You have evidently forgotten that I told you Sir Roderick had no doubt of the Spynie rock belonging to the Old Red; he saw the specimens, both *Stagonolepis* and *Telerpeton*; was convinced of the identity of the two beds; and he knows the country well."[1] Here Murchison is being referred to as *the* authority in the field. Roughly at the same time, a third edition of Anderson's *Guide to the Highlands and Islands* was published and, although the geology of Moray was now relegated to a long footnote, that was an entirely revised, updated account by Alexander Robertson, one of Gordon's friends. At Cummingston, near Burghead, the footprints of some as yet unidentified four-legged creature had been found on a slab of rock (Figure 10B). That had been in 1850, but the creature had not been identified by Captain Brickenden, the discoverer of the footprints. John Martin, the Museum's curator, wrote an article on the "Northern Drift"[2] and M. E. Grant

---

[1]ALS. Mantell to Charles Lyell, 25 December 1851. APS.
[2]John Martin (1800–1881) was Schoolmaster at General Anderson's Free School from 1830

Map 6. Geological map of Moray by John Martin from Duff (1842)

Duff published a more general piece on "Morayshire" in the *Westminster Review* of January 1858. As his later pronouncements make clear, Gordon continued to keep a close watch on anything which concerned Moray, while Murchison, unprepared for the storm that was about to break, talked about the geology of Moray in his old style at the 1858 British Association meeting in Leeds.

In late August and early September 1858 Murchison was again in Scotland accompanied for some of the time by Charles Peach. We can follow his movements fairly closely with the help of yet another field notebook: M/N134 in the Library of the Geological Society of London (see Figure 11), the label of which reads

---

to 1836 and, from 1840, Curator of the Elgin Museum. He studied the natural history and geology of Moray in depth and was a close collaborator of George Gordon. Martin published a prize essay on the geology of Moray in 1837 as well as his article on the "Northern Drift."

Map 7. Nicol's geologic map of North-East Scotland

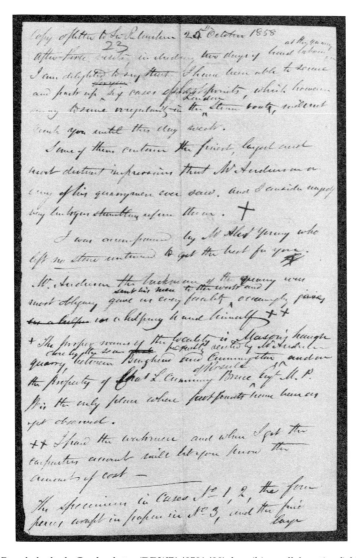

Fɪɢ. 9. Rough draft of a Gordon letter (DDW71/8581/28) describing collaborative field work.

Scotland
Skibo—Elgin & Reptiles
Perthshire—Kinnaird
Dr Anderson—Dura Den

It is important to notice at this point the relationship of notebook M/N134 to the "journal" that was later derived from it. In the first place, pp. 58–72 and 74–75 of the notebook, though covering the same journey from Inverness to Elgin, are not used in the journal at

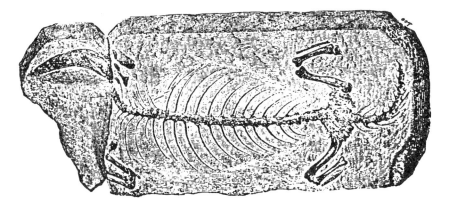

FIG. 10A. *Telerpeton* illustrated in Mantell (1852)

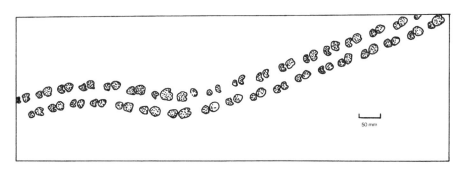

FIG. 10B. Reptile footprints from Brickenden (1852)

all for a reason that will become clear later. The suppressed notebook pages constitute a more accurate account of what Murchison
actually did than pp. 140–44 of the journal. In the journal, which it
will be remembered was being written by a secretary, Murchison
left Inverness on p. 140 and by p. 144 was in Gordon Castle, from
whence he was driven about by Prince Edward of Saxe-Weimar, the
Duke of Gordon being absent in Edinburgh. The abbreviated journal account gives more the impression of a social jaunt than a serious geological enquiry. It must also be understood that these
journal pages are not final manuscript copy but simply notes designed to establish the chronology, and the names of people and
places, possibly in preparation for the later writing of a book. They
are not pages that could actually be used in a book, being much too
disjointed. Furthermore, in 1868 the elderly Murchison permitted
himself interpolations that have no equivalent in the notebook, as
when he says: "No Director-General will ever meet with a more able
Quartermaster General than George Gordon." Here, then, is one of
the moments referred to in our preface when dependence on the

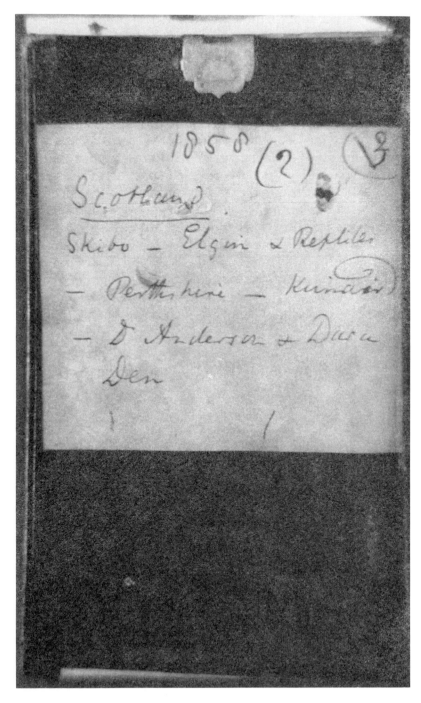

FIG. 11. Murchison's field notebook M/N 134

journal by itself would be a mistake. Murchison, incidentally, is still using his earlier tools: Cunningham's *Map of Scotland*, his—Cunningham's—1839 Memoir in the *Transactions* of the Highland & Agricultural Society of Scotland, and Anderson's *Guide,* probably the most recent edition. Later, an exasperated Gordon would comment scathingly on Murchison's dependence on the latter.

It is at this point—in September 1858—that our narrative connects with the correspondence reproduced in Part Two, Gordon's communications to the Geological Society in 1832 seem to have proved unnoticed but now Falconer gives Murchison a letter of introduction to Gordon. Interestingly it had been Falconer who had assessed Malcolmson's 1838 address to the Geological Society—the one that had been reported in *The Athenaeum.* Malcolmson's manuscript carried interpolations in Falconer's handwriting. (It will be remembered that Falconer, like Gordon and Malcolmson, had been in Edinburgh during the eighteen twenties). At the School of Mines in London, T. H. Huxley had begun the long task of analyzing the fossil fragments of *Stagonolepis robertsoni,* (see Figure 12) with the intention of giving a paper on it at the Geological Society that winter. Murchison

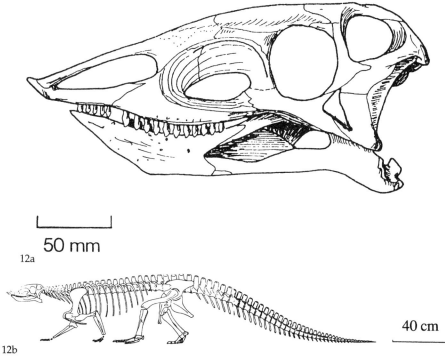

50 mm

12a

40 cm

12b

FIG. 12. *Stagonolepis* from Benton and Walker (1985)

knew something was brewing; but what exactly? When he reached Elgin on 6 September, Gordon took him to the Museum. "A most satisfactory afternoon with the Rev. George Gordon of Birnie, Murchison recorded in his notebook. "Examined first the collection of P. Duff which I saw in 1840. Now much enriched; particularly by casts of vertebrate creatures from the sandstones N of Elgin."[3] Later, with Falconer's help, Murchison prevailed upon Duff to let Huxley borrow these specimens, which was what allowed Huxley to complete his first paper on *Stagonolepis robertsoni.* But when Murchison was with Gordon in Elgin, it was probably impossible for him, at that time, to appreciate the implications of what he was being shown.

From the Museum, Gordon took him to Spynie Castle, Spynie quarry (which he sketched), then inland a little to Findrassie ("the sandstone quarries in which the *Telerpeton* was found" (see Figure 13)) and finally to Lossiemouth. On the following day—7 September—they made a long, looping expedition to the Knock of Alves, Burghead and Cummingston, and finally to Covesea where they were able to walk along the beach at low tide, an exercise which in this part of the world always made Murchison feel secure about the dip and continuity of the Old Red Sandstone. Gordon was taking Murchison to places that had become important since 1840, whether at Murchison's request or on his own initiative. With a topographical map in front of us (see Map 4), we can notice that on the first day Murchison would have seen the Quarry Wood ridge from Spynie stretching away to the west, while on the second day he would have passed the west end of this ridge on his way to the Knock of Alves. Though he inspected a quarry at Newton, he did not walk along the ridge crest in 1858, as usual making inferences about what he did not see from what he did. Given the distance from Elgin, it is unlikely that they traveled on foot. Murchison was sixty six. He had probably hired some form of conveyance at the Gordon Arms where he was staying. On the third day, they went inland across the braes of Birnie, but apparently neither to Pluscarden or Scaat Craig. What Murchison chiefly wanted to understand was the relationship of the rocks in the immediate vicinity of Elgin. In his notebook he wrote: "It is clear to me that much of the cornstones of Elgin must get under some of the sandstones of the higher slopes lying to the N & NW of it while other cornstones overlie all the sandstones—as near Lossiemouth."[4] We think this and a few other similar entries show that Murchison was having difficulty with the interpretation

---

[3]Notebook M/N134 folio 52. The Geological Society of London.
[4]M/N 134. folio 67.

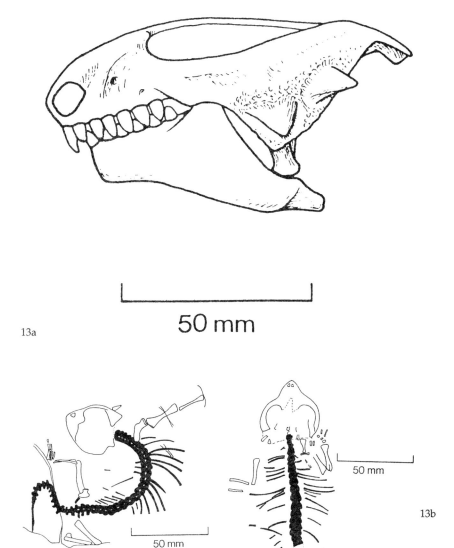

13a

13b

FIG. 13. Telerpeton from Benton and Walker (1985)

of the Elgin strata even before he understood the full significance of the reptiles.

Murchison's field notes included a "Grand Section of the Elgin County" (see Figure 14) which was reproduced as three figures in his 1858 paper (see Figures 15, 16 and 17). The section has a basal

FIG. 14. Grand sketch of Elgin section from M/N 134 f.72&73

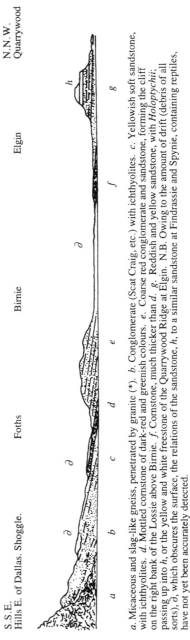

Section from the Crystalline Rocks across the Old Red Sandstone to the Yellow variety of the same at Elgin.
(Length about 11 miles.)

S.S.E.
Hills E. of Dallas. Shoggle.   Foths   Birnie   Elgin   N.N.W.
                                                                                      Quarrywood

FIG. 15. Elgin section from Murchison (1858)

a. Micaceous and slag-like gneiss, penetrated by granite (*). *b.* Conglomerate (Scat Craig, etc.) with ichthyolites. *c.* Yellowish soft sandstone, with ichthyolites. *d.* Mottled cornstone of dark-red and greenish colours. *e.* Coarse red conglomerate and sandstone, forming the cliff on the right bank of the Lossie above Birnie. *f.* Cornstone, much thicker than *d.* *g.* Reddish and yellow sandstone, with *Holoptychii;* passing up into *h,* or the yellow and white freestone of the Quarrywood Ridge at Elgin. N.B. Owing to the amount of drift (debris of all sorts), *∂,* which obscures the surface, the relations of the sandstone, *h,* to a similar sandstone at Findrassie and Spynie, containing reptiles, have not yet been accurately detected.

Section showing the position of the Reptiliferous Sandstone at Lossiemouth.
(Length about 3 miles.)

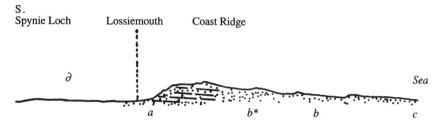

a. Laminated red and yellow sandstone (in the suburbs of Lossiemouth). *b** White and yellowish sandstone with Stagonolepis and Hyperodapedon. *b*. Yellowish sandstone, forming the east end of the Burgh Head, Coast Ridge. *c*. Overlying cornstone on the sea-shore. *∂*. Drift, covering the plain.

FIG. 16. Reptiliferous sandstone at Lossiemouth from Murchison (1858)

Section across the Coast Ridge at the Clashan Quarries.

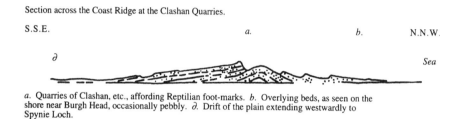

a. Quarries of Clashan, etc., affording Reptilian foot-marks. *b*. Overlying beds, as seen on the shore near Burgh Head, occasionally pebbly. *∂*. Drift of the plain extending westwardly to Spynie Loch.

FIG. 17. Coastal Ridge at Clashan quarries from Murchison (1858)

unconformity at Shoggle to the south and the strata dip consistently to the N.N.W at a gentle angle. To the south of Elgin, the section contains two horizons of cornstones (calcareous concretions) inter-bedded with red sandstones and conglomerates. Much of the section is obscured by drift but Murchison portrays the sequence as continuous and unbroken due to the consistent dip of the strata and their similarity to the Findhorn section. The ridge of Quarry Wood, according to Murchison, had red and yellow *Holoptychian* sand-stones at the base which passed upwards without a break into the yellow and white reptiliferous sandstones at the top. He noted that the location of the transition was obscured by drift but the unifor-mity of dip suggested to him that the succession was conformable. He correlated the sandstones of Quarry Wood with those of the coastal ridge at Lossiemouth (see Figure 16) and Clashan quarries (see Figure 17). He was able to do so by inferring a broad syncline beneath Spynie Loch on the basis of small-scale folding in the coastal ridge outcrops near Clashan. At Lossiemouth, Murchison noted that red beds of Old Red Sandstone character were con-

formably overlain by reptiliferous sandstones which in turn graded
up into siliceous cornstones along the coast. He noted that the
coastal cornstones were siliceous or cherty in nature, but included
them with the two calcareous cornstones south of Elgin as one of
three zones of cornstones. Thus, his section illustrates Murchison's
preference for stratigraphic continuity. Where the strata were obscured by drift, he assumed that the strata continued unbroken. The
importance of folding in the coastal ridge was minimized by inferring a broad, gentle syncline. Satisfied that his field work did not necessitate major revision of his earlier findings, Murchison left Elgin
on 9 September on his way to Gordon Castle, "after three mornings
work with my excellent, intelligent & most capable friend, Mr.
George Gordon."[5]

We now enter a brief but extremely interesting period extending
from the time that Murchison left Elgin to 15 December, when he
presented his paper on the geology of Moray to the Geological Society. Both men tried to make sense of what they had seen together
but their motives and methods were somewhat different. Murchison in his normal way chiefly wanted to establish the order of deposition around Elgin on exclusively stratigraphical grounds. His
notebooks show that when he went to the Spynie and Findrassie
quarries this was what interested him. This was what he always did;
he trusted the work that he himself had done. He had been to Spynie
and Findrassie, to Newton and the Knock of Alves, and to Cummingston and Covesea. This should have been enough, he thought,
to establish the actual succession of the strata, with their dip and
strike. Nowhere do we find him asking Gordon—"are there other
sites we ought to visit?" Of course one cannot tell with certainty
what they talked about, but, even if they had gone elsewhere,
Murchison would have been seeing with the same eyes, those of an
early nineteenth century stratigrapher.

While Murchison was still in Elgin, Gordon asked him about Malcolmson's 1839 paper for the Geological Society, since it became obvious during the few days they spent together that Malcolmson had
already investigated some of the problems that were now perplexing Murchison. Why had Murchison not ensured that the paper was
published? Had he blocked its publication? The correspondence reproduced in Part Two shows that Murchison must have promised
to look into the matter when he was back in London, but it also
shows that Gordon was distinctly dissatisfied by the neglect of Malcolmson's research. We have already referred to this unpublished

---

[5]M/N 134. folio 73.

Malcolmson paper in Section 3, where we demonstrated that Gordon continued their shared work from 1840 up to 1859. We now take a second look at the way Gordon supplemented what he and Malcolmson had done in the late eighteen thirties. Given Murchison's unsatisfactory response to his questions about Malcolmson, Gordon resolved to write a paper of his own which would set the record straight and also be an act of homage to his friend. He already had a manuscript copy of the 1838 paper; from Stables he obtained a draft of the 1839 paper. He may also have consulted Falconer. When he studied fossil fishes, he had simply let the research continue at a slow but steady pace, sharing his findings with local friends, as they did with him, until the Museum had an excellent collection of specimens. In the case of the geology of the region, he decided to re-check everything Malcolmson had written by revisiting all the sites they had together visited twenty years earlier. This decision was obviously a result of Murchison's short stay in Elgin. In other words, when Gordon wrote the article entitled "On the Geology of the Lower and Northern Part of the Province of Moray" he did not merely reproduce what Malcolmson had said. In the article as published in the *Edinburgh New Philosophical Journal*, the sections lifted from the Stables' draft are in inverted commas, while Gordon's, of course, are not. To notice what Gordon said on his own account as a result of checking Malcolmson's text is instructive. In October 1858, after Murchison's visit to Elgin but before the meeting of the Geological Society on 15 December 1858, Gordon went to all the old sites, which for the most part were places Murchison had never visited. We can assume there was anger or frustration on Gordon's part since (at least in the extant part of the correspondence) he refrained from telling Murchison what he was doing in the field, merely announcing that he was writing a paper of his own that would incorporate Malcolmson's results. In short, some thirty years after he had done his geology at Edinburgh, the Minister at Birnie decided to take matters into his own hands.

One has only to read a few of Gordon's interpolations, and compare them with the questioning, uncertain, or overly dogmatic tone of Murchison's letters in 1858, to be aware that he had made a determined effort to set right in his own mind the very matters that had perplexed Murchison.

In three of the river beds, and in some intermediate localities which we have lately revisited, the rocks that underlie the Old Red Sandstone are found of the same general aspect. They are for the most part gneissose, but contrast greatly with the coarse-grained, well-defined gneiss, as seen in the boulders scattered over the country, and occasionally met with in the very ravines that expose the gneissose strata. They lie unconformably under the Old Red Sandstone strata, having their dip to the east of north, and at a much more acute angle than the overlying conglomerates which dip to the west, at a low angle, of the same cardinal point. Their beds are much fractured, and present many cleavage surfaces, to such an extent, and so regular, that in some places these surfaces are apt to be taken as indications of the lie of the strata.[6]

How long Gordon spent on what was in effect a new survey of this territory cannot be determined exactly, except to say that it had been completed by 30 November. The passage quoted above shows, and Gordon elsewhere states, that this territory extended south to the Grampian watershed, that the huge amount of drift showed that the action of water, and perhaps ice, had been considerable, and that to observe the relationship of strata, which was what Murchison had wanted, one might have to go inland.

Lower Craigellachie, the hill of Conrack, and the ravines which radiate from the village of Rothes, are good points for the examination of these rocks on the eastern part of the province of Moray. Conrack is interesting on account of the extent and varieties of quartz that are found there. Although of considerable thickness, so as to form the body of the strata, yet it resembles that generally found in veins. It is white, with iron-shot streaks, and at times coarsely fibrous or radiating. Drusy cavities, lined with rock crystal and amethystine quartz, are also found in it. In a ravine, running westward from the north end of the Glen of Rothes, large sections of this portion of Silurian strata are to be seen.[7]

Obviously Gordon is starting his survey by identifying the district's oldest rocks. It must be kept in mind that this was neither a casual, dilettante exercise, nor exclusively an attempt to rescue the research of a close friend, Malcolmson; Gordon had just spent three days with Murchison, listening to his talk and observing the way he worked. In the same way as he had resolved to know everything possible about the fossil fishes of his part of Scotland, so now he resolved to come fully to terms with its geological problems. One would give a great deal to have a record of their September conversations, and of Gordon's thoughts when he returned to the Manse. But perhaps one can guess at the quality of their talk by noticing the firmness of Gordon's reaction to it, as in the following passage:

---

[6]Gordon, George. 1859, p. 26.
[7]Ibid., 27

On the Lossie, and in the ravines through which its tributaries, the Shoggle and the Lenock burns flow, from five to eight miles south from Elgin, these strata are best developed, and exhibit large sections well adapted for their study. On the west side, and above the fall of water at the line of the Shoggle, the overlying conglomerate is seen resting unconformably. Some hundred yards further down the stream, on the same side, the upper surfaces of some of the thinner strata are recently laid bare, so that not only the dip but the flexures, and the apparent effects of side pressure, are distinctly seen over a considerable space. The upper beds of this series on the Lossie, as at the Dun Cow's Loup, are of a hard, compact flinty nature. As we descend they become less quartzy and more gneissose; the latter being their prevailing character. Some of the strata, however, are coarsely micaceous. Veins of white quartz cut across the strata at all angles. Pluscarden Hill presents the same series of strata, and with the same dip as seen on the Lossie. Beginning at the east end we have the flinty, and, occasionally, brecciated rock as the uppermost beds. Then they pass into those that have a gneissose appearance. In this hill, and nearer the Priory, a second set of beds present the quartzy appearance again, which in its turn gives place to the gneissose character. Immediately north of the Priory a bed of mica-slate of no great thickness crops out. The strata that underlie the well-developed Old Red Sandstones of the Findhorn begin at Sluie. Lithologically, these rocks (Silurian?) on the Findhorn, from Sluie to Logie, do not differ from those of the Lossie. They contain, however, many granitic veins which are absent from the former locality, and their dip is not so uniform and distinct. Immediately above the junction of the two formations at Sluie the older rock seems, from some under pressure, to have assumed the saddle-shaped form, the beds dipping in opposite directions at no great distance asunder. In no part of this district have the granitic or quartz veins been seen to enter the overlying sandstones. Hence, as Dr Malcolmson suggests, it may be inferred that these sandstone beds have received their present inclination from forces that bore upon them from a greater depth than has generally been alleged.[8]

Murchison had said that the Elgin cornstone "must get under some of the sandstone." This loose expression shows that he had not been able to work out the relationship of the strata. Gordon deals with this as tactfully as possible by referring to different dispositions of strata in different places. More importantly, he refers to sites that Murchison either did not know at all, or did not know very well: Scaat Craig, Bishop's Mill and Quarry Wood. These places he already appreciated were problematic.

Although it be a question not to be settled by names or numbers, yet, following Messrs. Sedgwick and Murchison, and the late Mr. Robertson of In-

---

[8]Ibid., 27–28

verugie, we would here venture to state an opinion long held, that it is far more probable that the silicious sandstones of Bishopmill and Quarry-wood do rest on the cornstones of Elgin. The strike and the dip of the beds clearly indicate this position, which is one not the less likely, from the absence of any apparent disturbance or fault from trap dykes or otherwise, to alter the appearance of the order in which we suppose them to have been deposited. The beds that lie under the cornstones of Cothall have few visible representatives in the neighborhood of Elgin save at Scaat Craig, while the silicious strata above the cornstone so largely developed in the eastern parts of the county may be represented in the west by the sand-stones of Moy, and, possibly, of Boath, near Nairn.

The first, the most extensive and valuable escarpment of these silicious sandstones, lies immediately to the north-west of the town of Elgin. It runs from Bishopmill westward, continuously for four or five miles, to the moor at Alves. The joints, false bedding, and cleavage of the strata are apt to mis-lead the unpracticed eye. But wherever the real surfaces of the strata appear, and are distinctly seen, they take the usual dip at a low angle a little to the west of north. This large and elevated ridge or escarpment, in its various beds, exhibits every degree of consistence, from the hard, pebbly, quartzy sandstones of the Millstone quarry, to the friable yellow beds of Bishopmill; and of colour, from the dull yellow brown of the latter place, to the rosy tints of the quarry at the foot of the Knock of Alves.[9]

What must Murchison have thought, one wonders, when he read the sentence: "The joints, false bedding, and the cleavage of the strata are apt to mislead the unpracticed eye." When he had sketched Spynie quarry in his notebook, he had written, "There is here no case . . . for my zealous friend Mr. George Gordon." Thus we see the two geologists at the exact point at which a difference of interpretation occurred.

Two large isolated masses, apparently identical with the sandstone of Quarrywood, and which we believe are *in situ,* are to be met with the one on the south-east flank of the Hill of Pluscarden, the other at Rinninver, about three miles south-west from the Pluscarden Priory. They have been both worked for building-stone. It were curious to ascertain the cause of their isolation from the great deposit of the same nature. The supposition of a vast and repeated denundation going on during the depositing of the sandstones of Moray seems a probable explanation.[10]

Here, again, Gordon is going to the heart of a problem that was to perplex geologists for the rest of the century. These isolated

[9]Ibid., 42–43.
[10]Ibid., 43–44.

masses or outcrops not only seemed anomalous in themselves and would remain so until explained; they would also, some of them, soon be perceived as important because of their fossil content.

The fossils that Gordon knew about in October 1858 were the fish fossils already discussed, the *Telerpeton elginense* that had been described by Mantell, and the *Stagonolepis robertsoni* that Agassiz had said was a fish. Other fossils and other casts had been found since then but not securely identified. Duff had some; the Museum had others; Gordon had his own collection. Might these fossils shed light on the stratigraphical problems that still perplexed Murchison? With those problems clearly in mind, Gordon wrote the following fascinating and challenging paragraphs.

Ever since the discovery of the *Telerpeton Elginense,* Mantell (*Leptopleurony lacertinum,* Owen), a strong suspicion has arisen, and prevails with many, that between the *Holoptychius*-bearing beds of the above-mentioned ridge, and the sandstone of Spynie, there must here exist a wide hiatus or blank in the great geological scale of the earth's crust, filled up in other areas of its surface by the whole of the Carboniferous and Permian, and no small portion of the Triassic, formations. It may be so. Still, this we affirm, that, so far as has yet appeared from repeated and careful inspection of the strata where accessible, they conform so closely in their strike, dip, position, and lithological character, as would, in the absence of these reptilian remains, have induced the most practised observers to infer, with little doubt or hesitation, that the sandstones of Spynie, Lossiemouth, and Cummingston, were but the upper beds of one and the same formation as the sandstones of Bishopmill, Quarrywood, and Alves. Another circumstance, which certainly tends to show at the same time a marked difference in the age of these beds is, that, so far as has yet been ascertained, as soon as the reptilian remains appear, those of fishes disappear,—no impression of *Holoptychius,* or, indeed, any other organism, has hitherto been found associated with them.

About a couple of years ago, lying on the surface of the Bishopmill silicious sandstones, near the east entrance to Findrassie House, and among the debris of a pit opened up for road material, Alexander Young, Esq., Fleurs, found a most interesting and valuable slab containing casts of bones and scales. He presented it to the Elgin Museum, where it is now to be seen. Although not actually taken from the living rock, the lithological features of the stone, and the freshness of its angles, lead us to infer that it had not travelled far from its original bed. The scales seem to be generically, if not specifically, the same as those of the specimen from Lossiemouth, which Agassiz considered to belong to a fish, and to which he gave the name of *Stagonolepis Robertsoni.* The casts, however, of the bones, which lie in such close juxta-position to the scales as to leave no doubt of their belonging to the same animal, give good ground, we think, for raising it to the higher class of reptiles. The surface of the slab is about eighteen inches square. One bone measures three inches across the two condyles, and the scales $2^1/_4$ inches by $1^1/_2$ inches. But a verbal description here is of comparatively little use. We trust that at least a cast of this slab, and that other fossils from this now so interesting a locality, will be sent to Aberdeen, by the time the British Association meets in that city, for the inspection of those able to decide as to the character and position which these animals should have assigned to them among the vertebrata. Dr Taylor, H.E.I.C.S., at the same locality, subsequently found some casts of vertebrae, which are now in the cabinet of Mr. Duff.[11]

Murchison had an essentially static view of any terrain, regarding it in the present—as *he* saw it—and not bothering how its characteristics developed. Gordon understood that the region was unstable, that the sea, the action of ice, and the north-flowing torrents of what are now the Findhorn, Nairn, Lossie and Spey had all, over a long period of time, conspired to render the region unusually complex, especially because of the so-called drift which prevented the relationship of the strata from being clearly seen. He also understood that the drift itself had to be explained. Furthermore, his earlier, painstaking work on fossil fishes meant that he was bound to be skeptical about Murchison's attempt to use them for elementary stratigraphical purposes. Moray was not Caithness. On the crucial question of the age of the rocks, however, the two men continued to be in agreement.

In 1995 the reader of this book who wishes to understand the problems facing Gordon and Murchison will naturally look at modern geological maps of Moray and Elgin (see Maps 8 and 9), and perhaps also at the first productions of the Geological Survey of Scotland. That reader will see at a glance what Murchison could not see—the repetition of stratigraphy by faulting. We particularly draw attention to the outcrops around Spynie, Findrassie and Lossiemouth, though the latter is now mostly concealed by twentieth-century house-building. Look also at the complexity of the Quarry Wood ridge and imagine what Murchison, who may never have been up there, would have seen when he examined it from any direction. Of course the answer is: not much. Notice also how easy it would have been to assume that Cummingston and Covesea were united by the same type of sandstone, remembering that Murchison spent no more than part of an afternoon inspecting the coast between Burghead and Lossiemouth. On the modern maps the faults are marked with bold dotted lines and their posi-

---

[11]Ibid., 44–45.

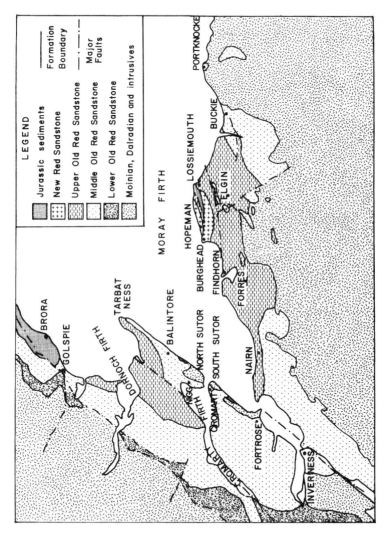

LEGEND

Jurassic sediments
New Red Sandstone
Upper Old Red Sandstone
Middle Old Red Sandstone
Lower Old Red Sandstone
Moinian, Dalradian and intrusives

Formation Boundary
Major Faults

MAP 8. Modern geologic map of Moray

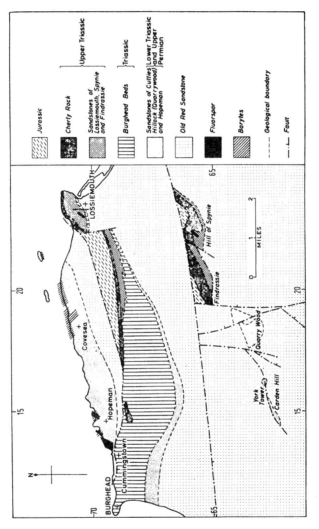

MAP 9. Modern geologic map of Elgin

tion largely inferred by the repetition of stratigraphy. To have detected them in the late eighteen fifties before the stratigraphy had been worked out would have been well nigh impossible. To have distinguished between the different types of largely unfossiliferous sandstone would have been difficult. Murchison would undoubtedly have had to spend more than a few nights at the Gordon Arms in order to do so.

By November the situation had become serious for Murchison because his notes from the Elgin visit were not adequate even for his immediate purposes—the paper he was to deliver in December. In the important letter of 27 November 1858 we find him saying:

My opponents seeing the very high character of the Reptilian remains (the Findrassie & Lossie Mouth ridges contain the same things) may contend that these white sandstones are of Jurassic Oolitic? age & have been deposited on flat or slightly inclined bed of Old Red with *Holoptychii*.[12]

One can guess that Gordon would not have liked this kind of talk. Opponents? It was much more in his nature to let the evidence speak for itself than to engage in controversy. But Murchison, clearly worried, continued:

I shall still hold to the white & yellow ridges of Elgin on the North & of Burg Head—Causie & Lossiemouth in the North (*for they* are now clearly *identified*) being upper portions of the Old Red, provided only the beds with *Holoptychius* fairly pass up into them. I also hold to this belief, because it seemed clear to me that the cornstones marked the Lossie Mouth beds.

There are however people who may contend, that the whole series of cornstones belong to an *overlying group* of secondary age, reposing on the undulations of Old Red with *Holoptychius*—that old [fraction] not extending northwards beyond the Elgin ridge & the younger deposits of like coloured Sandstone & cornstone having been accumulated thereon.[13]

Murchison's letter elicited a stiff reply from Gordon, who was now on top of his subject and one can guess did not appreciate Murchison's vacillations, inaccuracies and failures of memory.

I think we must admit that the Findrassie (a portion of the Spynie hill) beds have been deposited subsequently to those of Bishopmill. And Bishopmill from its containing *Holoptychius* scales being unquestionably Old Red,

---

[12]ALS. Murchison to Gordon. DDW 71/858/40
[13]Ibid.

the question is narrowed to—What is the age of all the cornstones & sandstones to the North of Findrassie from Burghead on the west, to Lossiemouth and Spynie on the east. It is simply the question of the reptiles that has raised the question, as prior to this (their lithographical character dip and strike agreeing with those in which the *Holoptychius* scales are found) we were led to regard them as Old Red too; and even now there seems no other element than these Reptiles that makes one hesitate as to this same conclusion.[14]

Tellingly, and quite understandably, he disagrees with what Murchison is still saying about fossil fishes.

I consider that the lowest fish beds (nodular) such as we have at Tynat & Dipple on the east side of this district, & at Lethen Bar & Clune on the west—if ever deposited have been afterwards swept away from the central part of Moray—particularly at Scat Craig and the Findhorn where the fossiliferous beds lie near to the crystalline rocks.[15]

Here once again one sees Gordon being more imaginative than Murchison. In his study of fossil fishes, he had obviously pondered the question of distribution. Why were there fossil fishes in some parts of Moray and not in others? The difference between the two men became more and more clear. Whereas Murchison invariably thought of fossils as evidence of stratigraphical continuity, Gordon understood that they would sometimes provide evidence of geologic events that had created discontinuity.

We have been suggesting to this point what indeed must be perfectly obvious, that Gordon's article for the *Edinburgh New Philosophical Journal* was a direct response to Murchison's September visit, even a riposte. We now turn to its effect. As it happened, both Murchison and Huxley spoke at the Geological Society on 15 December 1858. A few days before that meeting, Murchison tracked down the original 1839 Malcolmson article with Falconer's interpolations (see letter of 17 December 1858).[16] Gordon had arranged for the Edinburgh printer to send Murchison a proof copy of his article. Murchison therefore had both documents in hand before he gave his own talk, but whether he had time to take stock of the differences between them is doubtful. It's more likely that the letter of 5 December 1858 indicates what he was thinking about.[17] The bluster and agitation of this letter demonstrates the extent of his aware-

---

[14]ALS. Gordon to Murchison. DDW 71/858/41
[15]Ibid.
[16]ALS. Murchison to Gordon. DDW 71/858/46
[17]ALS. Murchison to Gordon. DDW 71/858/43

ness of his own vulnerability, so much so that he must have been hit hard when he later understood the implications of what Gordon had written. The British Association was to meet in Aberdeen in August 1859; Gordon and his friends already anticipated that the meeting would be the opportunity for further discussion of the geology of the North-East. Could Murchison get things right by then? In particular, would he have to revise the crucial sections of *Siluria* for the 1859 third edition? Huxley's article on the *Stagonolepis* had created a great stir. It was not a fish but a previously unknown extinct reptile. If this were so, what was the age of the rock in which it had been found? Murchison had a lot to think about. So had Huxley. Given his scarcely concealed dislike of Murchison he must have been greatly interested in Gordon's article.

Gordon had responded vigorously to Murchison's visit to Elgin in 1858, immediately setting to work to bring up to date and render more accurate the geological parts of Malcolmson's earlier article. Similarly, after hearing about Huxley's re-interpretation of *Stagonolepis,* he began to scour Moray for other reptilian fossils. He had been in correspondence with Huxley since the beginning of 1858, in December of that year sending Huxley the specimens that allowed him to continue the re-construction of *Stagonolepis.* Later that month, he despatched "four bits of Lossiemouth rock with remains of *Stagonolepis* in each."[18] By 30 December, Huxley could say: "even now I have in my hands the material for a fine monograph."[19] Now, in 1859, their correspondence became more urgent as, in his usual energetic and thorough manner, Gordon tried to put Huxley in the position to be able to talk about the Elgin reptiles (*Stagonolepis robertsoni* and *Telerpeton elginense*) at the British Association meetings in Aberdeen. "They (the specimens) belong to the Elgin Museum," wrote Gordon. "The Directors expect that, after you have made them by the chisel to reveal their contents, the specimens will be exhibited in Granite City; as they will be among the chief objects of geological interest, collected in the north for the forthcoming meeting there."[20] Once again Gordon was happily working in collaboration with an eminent scientist from the south.

That Spring there was an unexpected development. Gordon sent Huxley a new batch of fossils, probably in April, thinking that they were more *Stagonolepis* fragments; but Huxley immediately pronounced otherwise (see Figure 18).

---

[18]Michael Collie. Huxley at Work, Macmillan 1991, p. 104–105, DDW 71/858/49.
[19]Ibid. 106. DDW 71/858/51.
[20]Ibid. 107. DDW 71/859/25.

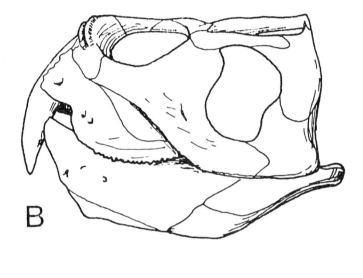

B

50 mm

18a

FIG. 18a and b. *Hyperodapedon gordoni* from Benton and Walker (1985) →

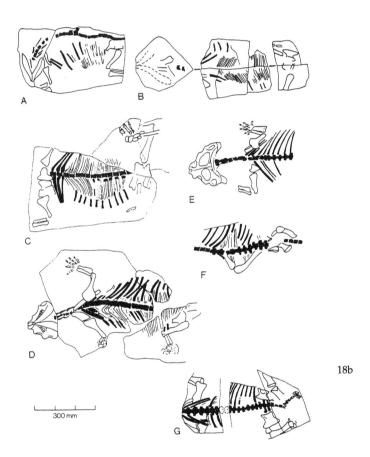

18b

I have put the fragments together & the hasty inspection which I have made of them is sufficient to enable me to say that the creature *is not Stagonolepis* & *is* one of the most remarkable anomalous Reptiles I have seen. Pray continue your search. The present specimen resolves many doubts I had about the first & every fragment is of value.

I shall be able to say nothing decided about the creature for some weeks. There is nothing for it but to sit down knife and chisel in hand, before the fossil for a week or two & fairly compel it to reveal its secrets by dint of scraping.[21]

This was followed a few weeks later by a second letter.

Your last package turns out to be one of the most interesting which has yet made its appearance. It is a *Lacertian* having some relations with *Placodus*, with *Rhynchosaurus*, with (alas for Sir Roderick!) unquestionably *Triassic* forms, and if you will forgive me for taking your name in vain, I call it *Hyperodapedon Gordoni*. The generic title is given on account of the pavement of teeth in its palate.[22]

This was a research breakthrough for Huxley, opening a large new field of enquiry. It was also a triumph for George Gordon, who was awarded an honorary degree by Marischal College later in the year, his work having been brought to the attention of the College authorities by Murchison.

As we said earlier, because David Oldroyd has already written a detailed, closely conceived book about Murchison's research in the Highlands—that is to say, in the North West—there is no need for us to recapitulate that complicated story, except to recollect the bitter controversy that had arisen over Murchison's descriptions and interpretations. When he began his work in the eighteen twenties, he had been a member of a small band of explorers and surveyors. Now, thirty years later, other people were more frequently studying his territory, assessing his earlier publications while making their own judgments about the terrain. Again, he committed himself to print *before* his problems had been resolved. Unwisely, as it seems in retrospect, he told his friend, Edward Forbes at St. Andrews, what he had intended to achieve during the summer of 1858: "My greatest objects however are traverses across Sutherland, Ross and Caithness to fix as I hope for ever the order herewith given and which I have already printed in my new edition of *Siluria* but not to be offered till November next.[23]

---

[21]Ibid. 107. DDW 71/859/23.
[22]Ibid. 108. DDW 71/859/26.
[23]Forbes Collection: St. Andrews, August 1858.

Murchison remained obstinant; he *would* continue to prepare his new geological map of Scotland, knowing that many controversial points of interpretation still had to be settled (see Map 10). Likewise, he pushed ahead with the third edition of *Siluria*, getting Gordon to revise and correct the Moray section despite all the talk in the School of Mines about the new reptiles, the impending British Association meeting in Aberdeen, and Charles Lyell's growing interest in Scotland. Lyell was someone whom Murchison respected and took seriously. How Huxley and Murchison managed to get on with each other during the first part of 1859 is a mystery, since each day's work made Huxley more confident about the age of the rock in which the Elgin reptiles must have been found. Each day must consequently have been more of a strain for Murchison, as he began to realise that their interpretations were utterly incompatible. At some point in 1859, they clarified their responsibilities for the Moray study. Huxley would attend exclusively to the palaeontology, Murchison exclusively to the geology. But this clarification or division of responsibility did not make life easier for Murchison; indeed the stress of having to admit to himself that he had been insufficiently thorough in September 1858 must have been considerable. He realized that he would have to re-visit Moray and in June he told Gordon that he will do so in the autumn:

Your last discovery of the *Hyperodapedon* of Huxley has so affected me, that I am a good deal shaken in my opinions as derived from the stratigraphical appearances & I propose to make a much longer foray into Morayshire than I did last year.

I will also reexamine the yellow Sandstones of Tain and Tarbet Ness in Ross-shire, which, as well as those of the Findhorn, seem to me unquestionably interbedded with the Old Red. But, (as I have put it in the note appended & trust you will return with the Slips) it is just possible that we may be able to find some signs of *transgression* between the fish beds of the Old Red & the Sandstones of the Elgin ridge.

I regret that I did not give a *whole day* to the Newton case i.e. in tracing the passage from the fish beds upwards. It certainly seemed to me that all the ridge there was one physical mass.

Possibly we may never, in your blind country, get any better evidence than we possess, but we must work hard.[24]

As evidence of what Murchison was thinking about, this important letter is nonetheless difficult to analyze. His decision to go to Tain and Tarbet Ness on the Black Isle indicates his determination

---

[24]ALS. Murchison to Gordon. DDW 71/859/30.

to demonstrate the continuity of sandstone strata from Caithness, where he feels professionally and intellectually secure, across the Moray Firth to Elgin, where he believes there ought to be a continuation of the same strata, notwithstanding the local disturbance. When he refers to the "Newton case," saying he felt sure the ridge constituted one physical mass, one sees that he has yet to become aware of the complexities of Quarry Wood. Will he inspect this area during his "much longer foray"? The word "case" presumably refers either to a discussion about interpretation that he had had with Gordon or to a letter we do not have. And what does he mean when he says he will revisit the Findhorn tract? Certainly he had crossed the Findhorn a few times on his way to Elgin and Aberdeen but he had never done any real field-work of the kind reported in Gordon's 1859 paper. Was he still convinced that his explorations in the eighteen twenties and 1840 had been adequate? This must be so. Though the correspondence is important and revealing, to be too severe about a research agenda in a private letter would be inappropriate; most researchers will have made such provisional lists. The question rather is: what will Murchison actually do when he gets to Moray this time? Problems have been identified; he intends to return to the old territory to check his former conclusions; this will be the visit, therefore, that most reveals what he does in the field. Will he solve the stratigraphical problems in his old manner? Or will he be forced to accept the reptilian finds as primary evidence? In this connection, the letter of 25 August 1859 is not encouraging. He was traveling with Professor Ramsay "who quite confirms all my views respecting the Old Succession." He could say this because he had been showing Ramsay Caithness and Rossshire. What would happen when they reached Nairn and Moray?

Murchison stayed the nights of 28 and 29 August at Cawdor Castle, where Stables showed him his collection of fossils and took him up the Burn of Achmeen south of the castle. George Gordon seems not to have been there; it had been left to Stables to guide Murchison to the sites Gordon had investigated in 1858. On 30 August Murchison enters in his notebook (M/N135): "leave Cawdor Castle in Mr. Stables' carriage and am driven up to the high country above Lethen and to the Findhorn valley." Now comes a series of notebook entries of extreme importance for the present narrative. Having been shown the exposed strata on the Findhorn—at the very same spot Gordon had already described—he wrote: "It is thus certain that the cornstone to the south of Elgin of which this is a western prolongation keeps under the yellow & light coloured

sandstone of Bishop's Mill and the Hospital Quarries. Q.E.D."
Murchison is assuming continuity between locations fifteen miles
apart. Preoccupied with continuity, he still wants to show that the
strata with fish fossils and those with reptilian remains are contiguous, and resolutely rejects the possibility of the presence of Triassic rock. Very tired after his exertions on the Black Isle, his
indisposition at Brahan Castle, and the series of long days, he sleeps
and dines at Fraser's Hotel in Forres, reaching Elgin on the following day, 1 September.

With Gordon and Ramsay he revisited Laveroch, Findrassie and
Spynie, sketching a cross-section in his notebook (see Figure 19).
"Now the strata in which the *Stagonolepis* occur are important in two
points of view." He measures the dip and strike of the strata he ob-
serves at Findrassie in exactly the same way as he had learned to do
in the eighteen twenties, connects them in his mind with the "same
descending section which is seen at Lossiemouth," and then, because of the angles measured in one or two places only, claims that
these strata "connect the reptilian beds with the mass of the Elgin
ridge in which *Holoptychii* occur." Of course, not everything that he
did on 1 September is entered into the notebook, so all one can say
is that there is no evidence that he went to Quarry Wood. Presumably they had walked or driven up on the east side of Laveroch, the
"Elgin ridge" being simply that ridge which rises above the town,
at that time without any housing on its southern slopes. At the end
of the day comes the aggressive notebook entry (toned down in the
journal): "From these data it follows that there is not a scintilla of
evidence to show that the reptile beds can be of Triassic age." On
the contrary. "the reptile beds are a regular & consecutive sequence
of the *Holoptychian* beds." The next day, while Murchison was "confined to the house by a bilious attack," Gordon took Ramsay to
Lossiemouth, presumably by train, in order to confirm "the existence of cornstone overlying the Reptile beds." On the third or
fourth, Murchison went by himself to the Bishop Mill quarry, where
he once again studied the terrain by applying the method he had
utilized for thirty years. The notebook entry is perhaps crucial for
an understanding of Murchison's predicament. "NB The top of the
stone in the Bishop Mill quarry is some 60 or 80 feet higher than
the Findrassie wood and quarries and as the dip is very gentle viz.
5° NNW and the distance between the two not more than half a mile
it follows that the Bishop Mill stone with *Holoptychii* must be immediately & conformably under the stone of Findrassie." The cross
section which follows is of immense interest as an example of how

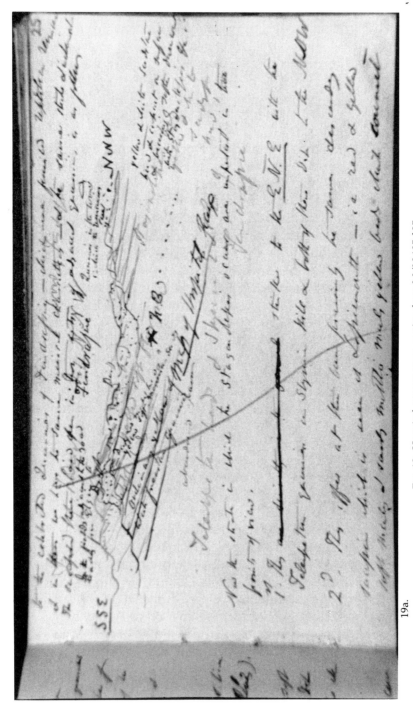

FIG. 19. Hospital quarry cross-section from M/N 135 f 25

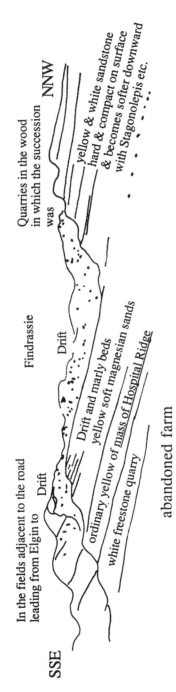

SSE

NNW

In the fields adjacent to the road
leading from Elgin to   Drift

Findrassie

Quarries in the wood
in which the succession
was

Drift

yellow & white sandstone
hard & compact on surface
& becomes softer downward
with Stagonolepis etc.

Drift and marly beds

yellow soft magnesian sands

ordinary yellow of nass of Hospital Ridge

white freestone quarry

abandoned farm

19b.

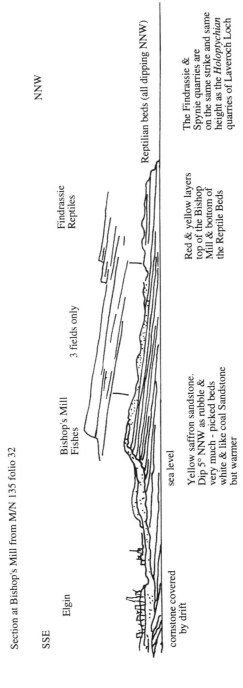

Section at Bishop's Mill from M/N 135 folio 32

SSE

NNW

Elgin

Bishop's Mill
Fishes

3 fields only

Findrassie
Reptiles

Reptilian beds (all dipping NNW)

sea level

cornstone covered
by drift

Yellow saffron sandstone.
Dip 5° NNW as rubble &
very much - picked beds
white & like coal Sandstone
but warmer

Red & yellow layers
top of the Bishop
Mill & bottom of
the Reptile Beds

The Findrassie &
Spynie quarries are
on the same strike and same
height as the *Holoptychian*
quarries of Laveroch Loch

FIG. 20. Bishop Mill quarry section from M/N 135 f.32

Murchison worked, since it demonstrates that he was prepared to assume (Figure 20) that strata lay conformably one on top of the other. A glance at a modern geological map shows that he was badly mistaken. The old methodology had failed the old practitioner. A sad business since, though he went to Lossiemouth the next day, this was in effect his final foray in the field.

Gordon reported the results of what had been an important four days in the field to his friend Hugh Falconer in a letter written in 1859 (DDW71/856/71). Part of his letter reads as follows:

On Wednesday week I met Sir Roderick Murchison & Mr. Ramsay at Forres & like the hunted fox we looked into every hole along the north side of the Pluscarden hill—from [   ] to Alves—clearly examined the passage of the red beds of Newton quarry into the white of the Knock; then inspected the Holoptychian bearing strata of the Hospital and Laverochloch quarries; thence to the Reptilian beds of Findrassie & Spynie back to the Holoptychian quarries of Bishop's Mill. Sir Roderick from an attack of diarhea was unable to go out with us next two days; but I lead Professor Ramsay over the ground on which Sir R. went last year.

Next day we took the Coast section from Burghead to Covesea, while Sir R. was able again narrowly to inspect the strike & dip of the reptilian & Holoptychian beds [where] they came nearest to each other at Findrassie & Bishopsmill. On the fourth day we were all on the Stotfield Skerries, etc.

The result is that, whatever palaeontologists may say, the Reptiles will be considered as firmly fixed in the Old Red as they were last year. The stratigraphical evidence is wholly in favour of this view, save only actual contacts.

The problem of identifying the unconformable contact between the Old and New Red Sandstones was to plague geologists investigating this area for the remainder of the century. It is greatly to Godon's credit that he was able to identify the crux of the problem at this early date.

5. THE BLACK ISLE (1859–67)

The Aberdeen meeting of the British Association in September 1859 was not a happy occasion for Murchison. There had probably never been such a gathering of geologists in the North. All the important people were there. James Nicol, professor at Aberdeen and already in conflict with Murchison about the North West, gave two papers. Huxley spoke about the Elgin reptiles. Lyell, whom Murchison had known for many years, was a significant presence; everyone knew and respected him. George Gordon and other directors of the Elgin Museum also attended. The North-East of Scotland had virtually been Murchison's private territory for several decades, but the discovery of reptilian remains near Elgin now brought the Moray Firth much more into the public domain. People wanted to hear about it, then see for themselves.

Murchison had been asked to contribute, not an ordinary specialist paper, but an evening address for what was anticipated to be a large audience. This address, published in the *Aberdeen Herald* 17 September 1859 and in the *Elgin Courant* 23 September 1859, but not in the *Proceedings* of the British Association, we reproduce in full as Appendix 5, because it is not widely known and difficult to obtain. A few points of interest may be noted here, however, because Murchison, perhaps thinking that it was his duty to be instructive, unwisely recapitulated his own experiences in the field, thus revealing and dwelling on his own attitudes, methods and concepts. First, he referred to his early research with Sedgwick with whom he had endeavored "to prove that the Old Red Sandstone of Scotland was simply the equivalent of that of England." Unfortunately this argument by analogy had proved unhelpful in the interpretation of specific geologic problems in the North of Scotland. Second, he recollected the instruction he had received in the field from William Smith from whom he had come to understand the "invaluable and truthful doctrine" that "each sedimentary formation was characterized by organic remains peculiar to it, and that there existed a regular order of superposition from the older to the younger strata." The comparison of Yorkshire, where he had worked with Smith, and Scotland he called "an accurate approximation." Third, he described his work with Ramsay in the North-West in 1859, claiming that Ramsay, "as honest as he is skilful in field-work," confirmed his idea that igneous intrusions "though productive of local distur-bance and partial alterations, caused no derangement of the order of succession beyond the limited areas in which such agency had been rife." In this he was badly mistaken. Fourth, he said that after he and Ramsay had revisited the Elgin region just before the Ab-erdeen meetings, they had come away "under the conviction that no stratigraphical evidence exists by which the strata containing reptiles can be separated from those charged with *Holoptychii* and other fossil fishes." All this was exceptionally injudicious. We obvi-ously do not need to dwell on Murchison's career-long retention of those relatively unsophisticated stratigraphical ideas that were in-troduced in Section 2 and which had proved unhelpful when he had to interpret complex local conditions such as those that existed in Moray. Any modern student of geology, studying, for example, Butler and Bell's pages on "Completion of the Cross-Section" in their *Interpretation of Geological Maps* would anticipate in *any* terrain the possible existence of faults and folds which have to be described and interpreted, not dismissed as mere exceptions to a vast general rule. Murchison's cross-sections of the eighteen sixties show, as we have seen, that he had not advanced beyond the concepts and mod-els of the eighteen twenties. He had failed to address specific ques-tions of geological interpretation in a way calculated to convince his colleagues. After the meeting they went to Moray to inspect the ter-ritory for themselves. The crush would include Charles Lyell, Pro-fessors Ramsay and Harkness, the Earl of Enniskillen and Sir William Jardine.

This does not mean that Murchison's Aberdeen address or the part he played in the British Association meetings should be dis-missed out of hand. Given the Royal Family's enthusiasm for Scot-land, it is unlikely that his subsequent visit to Balmoral in any way weakened his public request that geological survey work in Scot-land should be more strongly supported. Queen Victoria herself de-scribed the arrival from Aberdeen of "two huge omnibuses and carriages laden with *philosophers*" who were immediately treated to several hours of Highland Games which they watched, standing, from the Castle terrace. "We stood on the terrace," wrote the Queen, "the company near us, and the *savants*, also on either side of us."[1] Before the scientists returned to Aberdeen she had talked with Pro-fessor Owen, Sir David Brewster, Sir John Bowring, Mr. J. Roscoe and Sir John Ross, but of the members of the Association it was only

---

[1]Helps, 1873, pp. 124–6.

Professor Phillips, its secretary, and Murchison, who were invited
to stay the night. After dinner, if not at breakfast the next day, they
would have had Prince Albert's ear. Cynics might suggest that here
was Murchison again furthering his own career. Realists might
claim that the Queen wanted, not geological talk, but the news of
the Franklin rescue expedition that he had brought from London.
But geological colleagues presumably did not object to the Director
General's influence in high places, since his judgment about the
need for a separate and independent geological survey of Scotland
did in the end prevail.

Meanwhile, the response to Murchison's address varied. In the
Elgin Courant, Patrick Duff reported the speaker's opinion that there
existed "an unbroken series of beds of sandstone existing in Moray,
from the cornstones of cherty limestones downwards, which lie
conformably on each other, pass gradually into each other and attest epochs of tranquillity here, while the country all around bears
evident marks of violent convulsion and sudden change."[2] Granted
that his friend, Duff, was chiefly reporting for the benefit of local
readers the gist of the Aberdeen address, George Gordon may well
not have challenged these remarks, but his private thoughts could
have been akin to Uncle Vanya's in Chekhov's play of that name.
For all these years we in the provinces have been supporting a professor in distant St. Petersburg (read London) only to discover that
this after all is the extent of what he knows and thinks. Archibald
Geikie said dryly that Murchison had demonstrated he was not cut
out for public speaking. Unaware of this remark, Murchison had begun to confide in Geikie, who was helping him with a new geological map of Scotland (see Map 10). He had been "amazed," he said,
that his address had been overlooked by the British Association
"while all Nicol's trash is printed."[3] During the ensuing decade as
Murchison and Geikie spent more and more time together, the latter's loyalty and forbearance were remarkable.

It is virtually certain that T. H. Huxley is the key to the events of
the eighteen sixties. From the time that *Stagonolepis* was identified
as a reptile rather than a fish, it had been officially expected that, as
the School of Mines' comparative anatomist, he would contribute a
"Memoir"—that is, a book—to the series of memoirs put out by the
Geological Survey of Great Britain. When this eventually appeared
in the series, Huxley made a point of mentioning that he had been

---

[2]Duff, P., 12 September 1859, On the sandstone of the neighbourhood of Elgin: *Elgin Courant*.
[3]ALS. Murchison to Geikie, 4 July 1861. Geological Society of London: LDGSL 789/46.

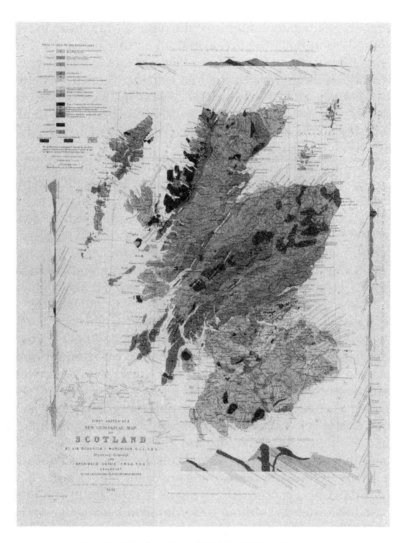

MAP 10. Murchison's and Geikie's 1861 geologic map

instructed to produce it by Murchison, who, as his superior, had made it one of his "duties." But "The Crocodilian Remains Found in the Elgin Sandstones, with Remarks on the Ichnites of Cummingstone" only appeared in 1877, six years after Murchison's death. No benefit would have accrued to Huxley from open hostility, since Murchison was still respected as a powerful figure in London even by those who appreciated that his time had passed. One remembers Featherstonhaugh's gentle note to Murchison in which he wished upon him "the greatest of all earthly blessings, a comfortable descent down the Inclined Plane of Life."[4] Others had the same kindly feelings towards the "old warrior." Somehow, between the publication of the third edition of *Siluria* in 1859 and the fourth in 1867, he did manage to keep going, and it is not to be doubted that Huxley watched him like a hawk. By this time Huxley's dislike of Murchison was intense. He believed him to be illogical, unprofessional and incompetent in the field, as well as pompous socially and blindly egotistical as a human being. He generally kept his opinions to himself, however, only occasionally indulging in indiscreet remarks to friends. If it is not possible to demonstrate conclusively that he delayed further publication on the Elgin reptiles until Murchison had put out the fourth edition of *Siluria*—playing cat and mouse, as it were—with his senior, there is nonetheless a strong likelihood that this is what happened. In 1867 Huxley published "On a New Specimen of *Telerpeton Elginense*" and in 1869 "On *Hyperodapedon*," ten years after Gordon had discovered it. Even for a person as busy, omnivorous and overworked as Huxley, this was a long delay. As he waited for Murchison to overcommit himself once again, his patience may have been spiced with an unworthy type of pleasure.

As long as Huxley delayed making public his description and interpretation of *Hyperodapedon*, Murchison could retain his opinions about the Old Red Sandstone around the Moray Firth, even though he knew that many of his friends and colleagues were skeptical. The period 1859 to 1867 was a breathing space for him as far as North-East Scotland was concerned. So it was that he could continue to work with Gordon even though Gordon had been active on Huxley's behalf. Like Murchison, Gordon remained unconvinced about the presence of Triassic rock in the region he knew so well. He was prepared to continue to support and encourage geological research around the Moray Firth if Murchison or anyone else thought it would be useful. For his part, Murchison seems to have believed

---

[4]ALS. Featherstonhaugh to Murchison, 20 September 1861. APS.

that, if indeed there were people in the North-East who wanted to continue the enquiry he had initiated, there was no reason for him to discourage them. It was at this point that the Rev. James Joass came on the scene.

After completing his studies at St. Andrews in 1854 (a university with which Gordon had connections), Joass became Minister at Edderton on the Black Isle, where he lived with his mother and sister in what is now the Old Manse. A talented draughtsman and cartoonist, Joass became the local geologist, naturalist and archaeologist *par excellence*. He later moved to Golspie where in the grounds of Dunrobin Castle he helped design, build and stock the little museum that is still open to visitors there. In the early eighteen sixties, Joass became deeply involved in the geology of Ross-shire and, in particular, the Black Isle. One could almost say that as Gordon had been Murchison's co-worker on the south shore of the Moray Firth during the fifties, so Joass was now co-opted into that role on the north shore. At issue, of course, was Murchison's contention about the continuity of the Old Red Sandstone from Caithness to Moray. Could this be observed on the Black Isle, the North Sutor and other parts of Ross-shire?

The substantial correspondence between Joass and Gordon, which is preserved in the Elgin Museum, documents the exploration of the Black Isle by Joass and his friend the Rev. George Campbell, in the early sixties. At home he had photographs of Owen and Murchison on his desk, and admiralty charts of the Moray Firth spread out on his table. Not knowing that his friend had cooled towards Murchison, he tried to persuade Gordon to join him in the field. Would he not cross the Firth to Cromarty or the North Sutor on the Heatherbell steamer? Before long Joass and Campbell found flagstones with "reptilian footprints." Excited by these first finds they made drawings, took photographs[5] and conducted experiments with various crustaceans to see what kind of marks they left on the wet sand. By the end of 1862 they had not only found more footprints but had also examined "the entire section from Nigg Ferry to Shandwick" on foot and by boat. Joass was certain that the footprints belonged to the Old Red Sandstone as shown by a drawing he made for Gordon's amusement (see Figure 21). Although Owen dismissed Joass's photographs of footprints as "nondescript," Murchison chose to write a flattering note of en-

---

[5]ALS. Gordon to Murchison 30 July 1862, DDW 71/862/12; Joass to Gordon, 3 October 1862, DDW 71/862/31; Joass to Gordon, 15 October 1862, DDW 71/862/33; Joass to Gordon 8 November 1862, DDW 71/862/38.

Fig. 21. Joass cartoon from DDW 71/863/7

couragement. "The discoveries are without any exception more
gratifying to me than any which of late years turned up in Scotland."[6] Later in the same letter he said that "the discoverers have
however brought to demonstration by fossil evidence what I have
always contended for that the yellow Reptilian sandstones of Elgin
have their true equivalent in the yellow sandstones of Tarbet-Ness
and Dornoch." Joass was doing exactly what Murchison wanted
and his letter made them continue their work through the greater
part of 1863. Whether Murchison really cared about what they were
doing is a matter for conjecture.

Joass continued his exploration of the Black Isle with undimin-
ished enthusiasm and energy. He sent Murchison a "tremendous
box of fossils." Had it arrived, he asked in January.[7] By February the
local laird's spare room, now a workshop, contained thirty fresh
specimens. In March Joass told Gordon: "I suppose I may now set
to work on my sections for Sir Roderick for I think I can do nothing
more to complete my notes."[8] Gordon was not so sure. Late in
March or early in April he crossed the Firth to spend a few days
with Joass and together they went round the North Sutor by boat at
low tide, making frequent beach landings hammers and other in-
struments in hand. It was on this occasion that they carefully
checked Joass's notes and cross-sections (see Map 11), compared
in detail what they knew about the rocks on the north and south

[6]ALS. Murchison to Joass, 23 December 1862. DDW 72/862/54.
[7]ALS. Joass to Gordon, 31 January 1863. DDW 71/863/10.
[8]ALS. Joass to Gordon, 9 March 1863. DDW 71/863/27.

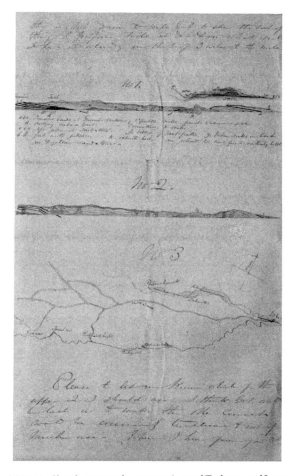

MAP 11. Sketch map and cross-sections of Tarbtness of Joass

shores of the Moray Firth, reviewed what had been published since 1859 such as Charles Moore's (1860) paper on the Elgin sandstones and Lyell's most recent pronouncements, and also took a grand general look at the development of geology in the North-East. Gordon's visit not only cemented their friendship but also gave Joass palpable encouragement. They kept in touch with each other, comparing notes on anything of geological interest. Joass's letter of 9 March 1864 shows that he well understood Gordon's qualities and achievements.

Speaking of our country as likely to rival Wiltshire in its profusion of ancient remains I must say that whatever it turns out will be but as a consequence of that spirit of intelligent research which originated on your side, & finds its exponent in your first-rate Museum—just as our chief interest as a Geological locality hinges upon the working out of your important stratigraphy.[9]

It was no doubt Gordon who introduced Joass to Harkness when Harkness became interested in fossil reptiles and footprints.[10]

While Joass, Campbell and Gordon were completing their work on the Black Isle (see Gordon and Joass, 1863, and Map 12), Murchison was in London leading for the first time a quieter life. He lived to be seventy nine, making one more visit to Scotland. He was naturally by no means as active as he had been in his middle years. Much of his time was devoted to two projects: the creation of the journal of his life based largely on his field notebooks, and the revision of *Siluria* for the fourth edition that was to appear in 1867. For both of these he had secretarial help, though unfortunately the handwriting has yet to be identified. One can imagine daily meetings in his study in which, as Murchison interpreted his notebooks and other documents for his scribe, he simultaneously reviewed the passage of his own life, looking back over almost four decades in the field. At the same time, we are reminded of the great personal importance to Murchison of the book called *Siluria*, the fourth edition of which appeared when he was seventy-five. We are also reminded of the way in which, over the years, he had conducted his research. As a field geologist he had been prepared, for better or worse, to strike out by himself, to enter little known territory with inadequate maps, and to endure tough journeys in order to implement what he had learned from his early mentors, making mistakes but quite frequently returning to correct them. As he grew older he recognized the need for a companion—Nicol, Peach, Gordon, Ramsay, Geikie—but even so he would still have to be reckoned as resolute in the field, even when his younger companions were physically capable of doing more than he could. In London, though, he was a different man—a man of clubs, associations and societies who believed in the accretion of knowledge because he had experienced it in the Royal Society, the Geological Society and the Royal Geographical Society. In this metropolitan life he was sensitive to his own and other people's reputations; people needed to get on with each other, reading and acknowledging each other's work, meeting to share their opinions, interpretations and judgments, and always talking openly about their work. This means, at least for Murchison, that work in the field was always tested in London, as the rock gave way to the word, and

---

[9]ALS. Joass to Gordon, 9 March 1864. DDW 71/864/14.
[10]Harkness, R. On the reptiliferous rocks and the footprint-bearing strata of the north-east of Scotland. *QJGS*. 20, 429–443.

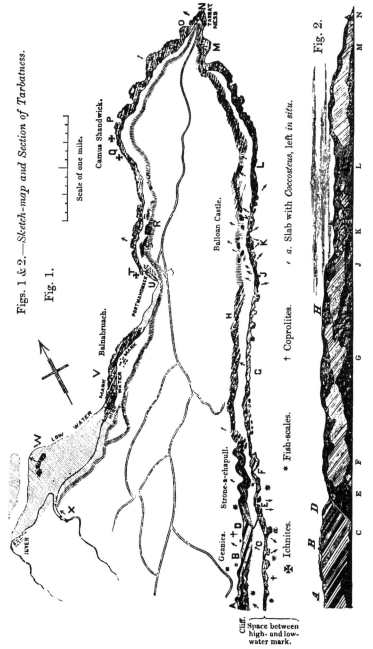

Figs. 1 & 2.—*Sketch-map and Section of Tarbatness.*

Fig. 1.

Fig. 2.

MAP 12. Geologic map and sections of Tarbat Ness from Gordon & Joass (1863)

it also meant that a book like *Siluria* bore the marks of the process by which knowledge was manufactured in London. As a writer, Murchison was not one to skip away to some rural retreat where he could formulate his ideas in isolation. On the contrary he had his readers constantly in mind; they were the other geologists, the other scientists with whom he had shared his whole career. If a line could be drawn between the dependence on other people's research that is recorded in the acknowledgment of it and the independence of mind that sometimes has to risk offending others for the sake of an original contribution to a field of study, one would have to say that Murchison located himself much more on the social than the independent side of that line. His independence had been in the field; in town he believed in inter-dependence.

In Chapter XI of the fourth edition of *Siluria* now entitled "Upper Sandstone of Moray, and Ross Triassic," and which is the only part of the book which concerns us here (i.e. pp. 262–268) Murchison re-shapes his old account in a startling manner. In the first place, it will be remembered that he had not been able to use his 1859 research findings in the 1859 edition of *Siluria* because the book was already in press by the time he went to Scotland. He therefore refers to the additional fish fossils he had obtained on that trip "through the kindness of Mr. Alexander Simpson."[11] There were specimens of *Pterichthys* (described by Agassiz), *Coccosteus* ("put together" by Hugh Miller), *Glyptolepis*, *Osteolepis* ("first described by Cuvier at my request"), *Cheiracanthus*, *Diplacanthus* and also *Dipterus* ("from the cabinet of Sir Philip Egerton"). He takes the occasion to correct a few mistakes in the first edition that had been carried through the second and third. By this time he knew much more about fossil fishes than he had in 1839, or even in 1859, and he had also closely considered Hugh Miller's *The Old Red Sandstone* (1841) and *The Footprints of the Creator* (1847). The re-writing of this section of *Siluria* is essentially an act of consolidation. His defenses had been breached; he therefore set about to repair them. Though he knows more about fossil fishes, and wishes to demonstrate the fact, his position on the geologist's use of fossil information has not changed. The geologist is interested in fossils only as the means to establish the continuity of strata.

Part of the revision of this part of *Siluria* takes the form of a long footnote that is so germane to the story being told here that it seems best to reproduce it in full.

------

[11]*Siluria* p. 262.

Already in 1828 Professor Sedgwick and myself united into one geological group the lower red conglomerate and sandstone, intervening cornstones, and the yellow fish-bearing sandstones of Elgin. We further showed that the fish bearing zone of Caithness was traceable to the south-east of Inverness as a thin course of shale though we there detected no Fishes in it. (Trans. Geol. Soc., 2nd ser., vol. iii, p. 147, 150 *et seq.*)

The remains of fishes were not discovered until nine or ten years afterwards,—first, I believe, by the zeal of the late Dr. J. Malcolmson. Lady Gordon Cumming, so well known to the readers of Agassiz and Hugh Miller, followed up these researches. Dr. Malcolmson, when on leave of absence from the East Indies, detected several Caithness Fishes at Clune and Lethen Bar in Nairnshire; and, having followed up his discoveries into Morayshire, he presented to the Geological Society a detailed memoir descriptive of these tracts in 1839. This paper was to be printed in the Transactions of that Society; but, as the author spoke doubtingly respecting the genera and species of his fossil fishes, the order for the publication of the memoir was deferred by the Council, until Agassiz, the great authority on Ichthyolites, should have determined the specific character of those fossils. In the meantime an abstract, giving the main features of the labours of Malcolmson, was published (Proceedings Geol. Soc. Lond. vol. iii, p. 341.); and shortly afterwards that accomplished man, having returned to India, fell victim to his zeal in pursuing researches in the jungles of the Bombay Presidency. Subsequently, in 1859, the substance of this memoir, with its illustrative sections, was published in the *Edinburgh New Phil. Journal* (January, 1859) by the Rev. G. Gordon, and in the *Quart. Journ. Geol. Soc. London,* vol. xv, p. 336.) Dr. Gordon has well illustrated all the fish-bearing strata around Elgin which belong to the Old Red Sandstone, as well as those overlying sandstones which I now refer to the age of the Trias, as will be presently explained.[12]

Most authors from time to time revise their work, so who will pass judgment on Murchison's decision here? Certainly he had intellectual reason enough for re-writing and re-casting the whole section even at the risk of expanding it and so disturbing other parts of the book, but he chose to tinker with what already existed, believing no doubt that the main line of his argument was secure, reptiles or not. A consequence of this decision is that his revisions seem more cosmetic than substantial. His paragraphs on fishes (pp. 262-266) deserve a monograph of their own, since they constitute not much more than an unanalyzed, uninterpreted list—an extended list, certainly, insofar as Murchison mentions the recent finds of Joass and Miller, but one which further exposed the author's lack of

---

[12]Ibid., 264.

curiosity about the fishes themselves. He makes no reference to Gordon's work on fossil fishes referred to in Section 3 above. Essentially Murchison is saying here that he had been right in the first place; that is, that the discovery of more fossil fishes confirmed his description of the Old Red Sandstone from Orkney to the Moray Firth. In the footnote, he says that the publication of Malcolmson's article was delayed until Agassiz had inspected the specimens, but he does not say whether or not Agassiz did so. He says that the "main features" were published in the *Proceedings* but he does not explain what bearing, if any, these "main features" had on his own work. He says that the "substance" of Malcolmson's paper was published by Gordon but he makes no reference to the significance, for him, of either Malcolmson's or Gordon's detailed research. In short, while this footnote might give the unsuspecting reader the assurance that Murchison was being punctilious, the author was in fact constructing a barricade with empty cardboard boxes. Anyone who checked the references that the note provides would quickly discover that Murchison had been economical with the truth, or else in his old age had simply lacked the energy to come to terms intellectually with developments in his field since 1858.

The section on fossil fishes which Murchison tinkered with while preparing the fourth edition for the press was merely the prelude to the much more important revision of the paragraphs which followed. In these Murchison in effect surrendered part of his ground, giving in to the claim that the rock in which the reptilian remains had been discovered could not be Devonian but had to be Triassic, just as he believed Huxley had always claimed. This revision as far as the story of Murchison's field work in North East Scotland is concerned, is its climax. We therefore reproduce the crucial paragraphs in full.

Whilst I adhere to that triple classification of the Old Red Sandstone of the North of Scotland which I correlated with the similar arrangement of the Devonian rocks of Devonshire and the Rhine, particularly where the series extends from the Ord of Caithness northwards into the Orkney and Shetland Islands, I have to announce that in respect to the age of the uppermost light-colored sandstones of Burgh Head and Lossie Mouth, south of Elgin and of Tarbet Ness (Ross-shire), I have now abandoned the suggestion of classing them with the Old Red or Devonian rocks. Stratigraphically considered, the strongest grounds indeed still exist to induce the field-geologist to connect these reptiliferous sandstones with the subjacent Old Red Sandstone, on which they repose conformably, as shown in my last edition. In the environs of Elgin I have three times examined these rocks, and on the last two occasions in company with my accomplished friend the Rev. G. Gordon of Birnie, and once when aided by Professor Ramsay. In advancing from the crystalline rocks on the south, and passing through the lower zones of Old Red Sandstone with numerous characteristic Ichthyolites, and from them through the chief yellow sandstones north of Elgin with their peculiar fossil Fishes, into the sandstones of the coast, I could detect no unconformity between all these beds with Ichthyolites and the strata of rather lighter colour, and containing concretionary cornstones, also like Old Red Cornstones, which extend to Burgh Head and Lossie Mouth. It is in these last-mentioned rocks that the remarkable Reptiles the *Telerpeton* (Mantell), the *Stagonolepis* (Agassiz), and the *Hyperodapedon* (Huxley) have been found.

Again, numerous footprints of Reptiles were found on the surfaces of these sandstones, whether between Burgh Head and Lossie Mouth in Elgin-shire or near Tarbet Ness in Ross-shire. In the last-mentioned district various geologists have confirmed the original observations of Professor Sedgwick and myself (1827), showing that all these strata seem to form a natural and conformable cover of the Old Red Sandstone and its Ichthyolites. Professor Ramsay and Harkness have sanctioned this view, as well as the Rev. G. Gordon, the Rev. J. M. Joass, and others who have examined that coast.

Stratigraphically, therefore, the evidence seemed almost conclusive; and as I had been assured by Professor Huxley that the Reptiles found in these rocks were unique and wholly distinct from any known Mesozoic forms, I was prepared to suggest that, inasmuch as they were purely of terrestrial or fluviatile origin, it might be that creatures of this high orga-nization were in existence when the earliest prolific flora flourished of which we have evidence.

But all such speculation has been set aside by a palaeontological dis-covery which Professor Huxley has made. A fossil Reptilian bone, con-taining teeth, found in the Keuper Sandstone at Coton End, south-east of Warwick, was recently brought to him by Mr. Lloyd, F.G.S., of that town. On inspecting this additional relic, Professor Huxley was unable to dis-tinguish it from the corresponding part of the Reptile from the Upper Elgin Sandstone (Lossie Mouth), which he had described and named *Hyperodapedon*.

To such fossil evidence as this the field-geologist must bow; and instead, therefore, of any longer connecting these reptiliferous sandstones of Elgin and Ross with the Old Red Sandstones beneath them, I willingly adopt the view established by such fossil evidence, and consider that these overly-ing sandstones and limestones are of Upper Triassic age, and must once have formed the natural base of those Liassic and Oolitic deposits of the north-east coast of Scotland which I described forty years ago.[13]

It is impossible not to be disappointed by this retraction, logical though it may appear to be. Having used fossil evidence through-

---

[13]Ibid., 266–267.

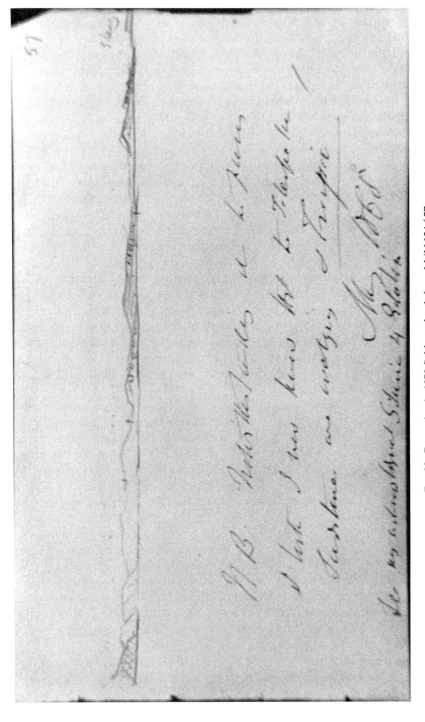

F_{IG}. 22. Retraction in 1858 field notebook from M/N 134 f.57

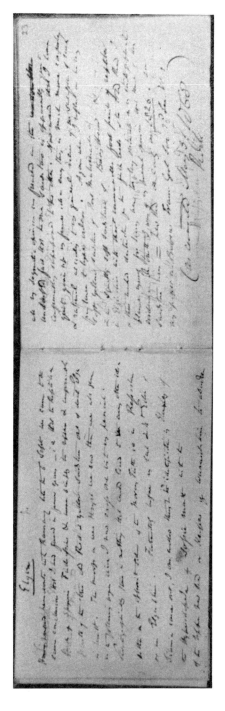

FIG. 23. Retraction in 1859 notebook from M/N 135 f. 22&23

out his whole career to establish the continuity of strata, there was no room for him to argue now that the Elgin reptiles were an exception. If *Telerpeton* had been found in Triassic rock, it had no place in his book and he therefore took it out. It will be remembered that while he was preparing this fourth edition for the press, he was also editing his notebooks into a "journal" with the help of a secretary. Over the notebook pages relating to his 1859 visit to the south coast of the Moray Firth he wrote the word "omit," thus creating an instance where the journal would be less reliable than the notebooks as a record of what Murchison actually did. On folio 57 of notebook M/W 134 he wrote "Notwithstanding the pains I took I now know that the *Telerpeton* Sandstones are . . . Triassic" and he dated this entry "May 1868. See my acknowledgement *Siluria* 4 Edition" (see Figure 22). On folio 22 of M/W 135, corrected on 3 May 1868, he said that it was Huxley's work that made him abandon all his "dogmatic opinion" (see Figure 23). Possibly one would say that these notebook corrections mark the end of a phase in the development of Victorian geology. If Huxley was right and Murchison wrong, there obviously remained a huge question still to be answered. How *could* there be Triassic rock above the pervasive Old Red Sandstone with no sign of the intervening Carboniferous and Permian strata? What geologic events had caused this major unconformity? It had never been Murchison's inclination to tackle such questions and now, at the end of his career, it was impossible for him to think of doing so.

6. A RETROSPECT

To this point we have attempted to make a modest contribution to the history of geology by isolating Murchison's work in the North-East of Scotland so that it could be described in sufficient detail for the problems and the tentative solutions to be seen clearly. Murchison probably tried to do too much. He depended too much on his *original* ideas about geology, failing to modify them sufficiently as his own experiences grew and the discipline developed during the first three quarters of the nineteenth century. While others were forming increasingly sophisticated theories about the causes of geological events, he held to his simple ideas about stratigraphy, even when these were being tested and modified by the survey of which he was Director General. *Siluria,* as well as his other major articles and books, together represent a historical period in which he was a pioneer and leader whose research was soon to be surpassed by others. Without wishing to detract from the achievement that these major publications represent, we now ask ourselves whether anything of a more detached, analytical kind can be said about his work in the North-East of Scotland, and in Moray particularly.

Of great interest is Benton and Walker's 1985 article "Palaeoecology, Taphonomy, and Dating of Permo-Triassic Reptiles from Elgin, North-East Scotland." The authors list five aims, two of which are to "assess the ecology of the late Triassic reptiles" and to provide "a summary of the Elgin reptiles and their environments."[1] This important article will be used as a starting point of our own reassessment, which will concentrate only on events, discoveries and interpretations during Murchison's lifetime. We attach importance to the term "Permo-Triassic" in Benton and Walker's title, since the relation of Triassic to other rocks in the Elgin region is open to interpretation even now. The authors distinguish between three formations: the Hopeman Sandstone Formation, the Cutties Hillock Sandstone Formation, and the Lossiemouth Formation, attributing a late Permian date only to the first of these. They are interested in the geology of the region principally in order to locate the reptiles and footprints that are the subject of their paper, not con-

---

[1]Benton, M.J., and Walker, A.D., 1985, p. 207.

113

cerning themselves with other rocks within the area bounded by Lossiemouth, Elgin, Pluscarden and Burghead. Despite this focus, they do address directly the problems that perplexed Murchison, particularly the problem of having to revise the geologic description of the Moray district to take account of the apparent discovery of Triassic rock where Murchison had thought there was only Old Red Sandstone.

The authors compare the Moray Firth to other parts of the globe, arguing, for example, that the Lossiemouth Sandstone Formation logically must be Triassic because "the *Stagonolepis* is closely similar to the *Aetosaurus* from the German Stubensandstein."[2] Here the palaeontologists are making the kind of comparison that is their bread and butter, in the process interposing geologic inferences that are not based on geologic evidence, which is interesting in the context of the conflict between geology and palaeontology which T. H. Huxley wrote about. Anyone with a serious interest in the geological and palaeontological problems of the Moray Firth will wish to be familiar with Benton and Walker's article. In order to clarify Murchison's position historically, we will discuss a number of points raised by their paper.

The first has to do with the Cutties Hillock Sandstone Formation, the term they allocated to a small area that, on their hand-drawn map, is west of Elgin on either side of the present Elgin-Forres road. This area is called Quarrywood on the Survey map and the Survey has named these same rocks Sandstones of Cutties Hillock (Quarrywood), equating them with the Hopeman Sandstones as Upper Permian to Lower Triassic. One of the most controversial sites on this ridge is Cutties Hillock quarry itself which can easily be reached on the old track or roadway which leads up directly from Elgin—to the west of Laveroch farm. After Murchison's death, this quarry became the focus of attention when other reptilian remains were discovered there. On that day in 1858 when Murchison looked west from Spynie, as he also did on other occasions, seeing a single, lightly wooded ridge, he was mistaken in thinking that the ridge represented a continuity of strata stretching from Spynie to the Knock of Alves. As we have seen, the assumption of continuity was his Achilles heel, tending to make him careless about detail when he had formed an overall picture of the relations of the principal strata. The detail he did not know about in this instance was a small, isolated outcrop of rock later classified as Triassic when reptilian remains were found in it. Benton and Walker say that "the

²Ibid., 229.

formation is seen to rest unconformably on Rosebrae beds of the Upper Old Red Sandstone in Rosebrae Quarry and York Tower Quarry,"[3] at the west end of the ridge to which Gordon could not have taken Murchison in the eighteen fifties and sixties because they had not yet been opened. Benton and Walker list the reptilian remains found at the Cutties Hillock site, most of which were described by E. T. Newton in 1893 and some of which were seen by Judd in 1884 and 1885. Without the quarries, it is doubtful that Murchison would have been able to detect the two different types of sandstone in the Quarrywood-Knock of Alves region, however much he had tramped about. Several decades after Murchison's visits geologists inspecting the Cutties Hillock Quarry had difficulty finding the unconformity between two formations. Had they been successful, they would still have been left with the problem of explaining the unconformity.

Though Murchison could not have known all the quarries to which Benton and Walker refer, he had been on several occasions to Newton and the Knock of Alves and to other places in the immediate vicinity of Newton. It is unlikely that he would have spoken with local farmers about what they had observed or found on the land. A sense of his own social position probably prevented him from seeking advice, for example, from the farmer and factor at Newton. In any case, of the three formations which Benton and Walker discuss, this was the least troublesome since he was unaware of any fossil remains that would make him anxious about the correct identification and designation of the rock. What we ourselves lack is a knowledge of on-site discussion between Gordon and Murchison. We cannot confidently say that Gordon, even when aware of his superior knowledge, ever took the conversational initiative, insisting, arguing, trying to convince. Possibly he often thought he had told visitors as much as they needed to know, or that he had answered their questions to the best of his ability. Besides, he was pre-disposed to accede to Murchison's ideas about the continuity of the Old Red. The question always has to be—at any given time—what evidence did they have at their disposal? Benton and Walker refer quite properly to specimens now in public collections. From Gordon's correspondence with both Huxley and Murchison we know he occasionally sent specimens—perhaps intrinsically important specimens—to friends, when there was little likelihood they would end up in museums. We also know from deletions from his letter drafts that he sometimes concealed, or did not reveal, a new

---

[3]Ibid., 216.

location from his metropolitan colleagues. One of these locations
was near Pluscarden. Would Murchison have done better had he
known everything Gordon knew and if he had worked out more attentively the implications of what Gordon said to him? The answer
is no. Starting out with presuppositions about uniformity, he was
both negligent about the *detail* of a terrain, and somewhat impervi-
ous to the findings of local geologists unless they supported his own
interpretation. Or put differently—although he visited Moray in
1826, 1827, 1840, 1858 and 1859, and knew that the region was geologically problematic, he never thoroughly surveyed the district, instead depending on transects and cross-sections, together with his
own and other people's supposed corroboration of them. Had he
known about the reptilian remains later to be found in the Cutties
Hillock Sandstone Formation, he would nonetheless have been on
the lookout for stratigraphic continuity. But if not engaged in ob-
jective survey, what *did* he do when he crossed and re-crossed increasingly familiar territory? We are suggesting that the answer can
only be found in his texts: his notebooks, his letters and his publications. The alliance of fieldwork and text of which we have given
an example has to be at the center of any history of geology.

We come now to the Hopeman Sandstone Formation, to which
Benton and Walker are prepared to give a late Permian date, maintaining at the same time that there exist no stratigraphic reasons for
connecting either the Triassic sandstone around Burghead, or the
Hopeman sandstones stretching eastward to Covesea, to those
other sandstones at Cutties Hillock. They say that "the tracks from
the coastal sandstones are believed to have been made by dicyn-
odont reptiles, but this is a group which ranges in time from the
middle of the Permian until late in the Triassic."[4] The proximity of
these two, maybe different types of sandstone—less than ten miles
apart—is not discussed. Instead the authors prefer to identify a formal stratigraphic unit which includes the reptile bearing beds and
a separate, different unit where tracks and footprints had been
found. Here, naturally enough, they are behaving as true palaeontologists. The locations and fossil types *are* different: what more
need be said? In Murchison's day only two sets of footprints had
been discussed, in an 1850 paper by Brickenden and the 1859 article by Huxley who, however, had never visited the place where
they were found. When Gordon's friends on the Black Isle began to
find other sets of footprints, Murchison's main interest was not
in the supposed dicynodonts themselves, but only in such evidence

---

[4]Ibid., 215.

as supported his hypothesis about the continuity of strata *under* the Moray Firth on a line, presumably, from Portmahomack to Lossiemouth. Here is the classic case of Murchison wishing to use fossil or imprint evidence *only* to establish the relationship of strata whose continuity could not be directly observed. Thus, in his letter to Gordon dated 24 April 1863, which is his response to the "map diagrams and Memoir" he had received from Joass, he says: "After all the sectional and stratigraphical evidences which have been obtained from the Tarbet Ness promontory (which I was always convinced would not explain the Elgin to Burg Head strata) it is very difficult to dissociate the fish and reptilian beds. They evidently belong to the same *consecutive geological series*."[5] In essence, Murchison appreciated the immense amount of work which Joass and his friends had done on his behalf, only to the extent that it bolstered his case against Huxley, as the revealing letter of 18 June 1863 shows. Joass, unlike the circumspect Gordon, was the enthusiastic amateur, examining the whole coastline of the North Sutor with care, making drawings and taking photographs, and even conducting experiments on how living animals left distinctive marks when they crossed loose sand. Then he left it to others to interpret the evidence he had assembled. In Murchison's case, neither the fossils nor the footprints set in motion a train of thought about the history and meaning of complex formations.

We come finally to the third sandstone formation discussed by Benton and Walker—the Lossiemouth Sandstone Formation, which they describe as "several small fault-bounded blocks at Lossiemouth, Spynie and Findrassie" perhaps overlying point bar deposits of the Burghead Sandstone Formation. They describe the Lossiemouth Sandstone Formation as "white, buff, yellow or pinkish," the rock being composed of "quartz, feldspar, and rare brownish chert and quartzite."[6] What the authors chiefly wish to establish and demonstrate is that a second assemblage of reptiles, distinct from the Cutties Hillock Sandstone Formation assemblage, were found in the Elgin district—not in the Old Red, nor in the 'Cherty Rock' which outcrops around both Lossiemouth and Spynie, but in the Lossiemouth Sandstone Formation only. As already stated, in Murchison's day these reptiles were *Stagonolepis*, *Hyperodapedon*, *Telerpeton* and, marginally, *Ornithosuchus longidens*. Many other specimens of these reptiles were discovered after 1871. Seven other extinct creatures have also been identified in the

---

[5]ALS Murchison to Gordon 24 April 1863, DDW 71/863/42.
[6]Benton and Walker, 1985, p. 218.

Lossiemouth sandstones from their partial remains, including a dinosaur. Perhaps in part because of the professional tension that existed between Murchison and Huxley, many years passed between Huxley's first paper on the *Stagonolepis* (1859) and his supposedly firm identification of *Hyperodapedon*-like creatures in Triassic rock elsewhere in the world (1869). This was a breathing space for Murchison. He had his back against the wall, but could avoid declaring himself until Huxley published. He was certainly perplexed by the *Stagonolepis* which Huxley had declared "crocodilian," but others were also perplexed. Lyell, Nicol, Symonds, Ramsay, Harkness and others all visited Moray to see for themselves. Only in the fourth edition of *Siluria* did the seventy-five year old Murchison deem it prudent to capitulate. As explained in the previous section, the signs of this capitulation are his letter to Gordon dated 24 January 1867, the deletion from his field notebooks of entries made in 1858 and 1859, and of course the revision of the relevant parts of *Siluria*.

Part of the problem was that, during the many years in which Murchison and Gordon held out for an Old Red Sandstone date for the rock in which reptilian fossil remains had been found, no one else did the close, painstaking survey work which might have resulted at least in the identification of the rock types to which Benton and Walker refer. Such work only began when the Geological Survey of Scotland reached the Moray Firth which was after Murchison's last visit. When what Huxley called the "big guns" of geology visited the Moray district at various times after the Aberdeen meeting of the British Association (as mentioned in the notes to the letter texts in Part Two of this book), they were for the most part primarily bent on establishing that the reptiliferous sandstone could not be Old Red but had to be Triassic. More sophisticated interpretations than this were not published until later in the century. How then shall we regard Murchison's own interpretation? In the eyes of his distinguished contemporaries he was undoubtedly mistaken. If identical or similar reptilian remains had been found in Triassic strata elsewhere, the strata in which they had been found near Elgin had to be Triassic. Here once again we see the willingness of the Victorian geologist to be persuaded by analogy. In the end Murchison, too, accepted that, despite the notebooks that he prized so highly, he must have been mistaken, and the crucial, perhaps embarrassing letter of 24 January 1867 shows his trying to deal with this. This letter was written when there was no likelihood of his being able to return to the Moray Firth. Evidently his main purpose was to render the fourth edition of *Siluria* safe from attack. He preferred to retreat and retain the respect of his contemporaries, rather than continue to skirmish on ground already lost—something which those fond of harping *ad nauseam* on Murchison's military metaphors of aggression might have in mind. But did he retreat too far?

Murchison was a Successionist. If it had been demonstrated that the fossil remains had been found in Triassic rock, he would have to accept the fact, even though he himself had not observed the contact points between various types of sandstone. Stratigraphy apart, he had no difficulty in thinking about different forms of life having been created at different times. It was by this means that he defused the time bomb of geology as it threatened social stability and religious belief. What would he have thought had he not been opposed to evolution, as his 1859 letter about Darwin shows him to have been? Presumably he would not have thought that the three or four types of reptile found in the Elgin sandstones had all evolved rapidly enough for the whole life span of each species to have been contained within a narrow band within the Triassic stratum. This cannot be right. Captured around Elgin is an interval in geologic time when these reptiles lived, but these reptiles probably existed before and after the time interval represented by the sandstones. Benton and Walker's Figure 8 shows a "scene at Elgin, north-east Scotland, in Lossiemouth Sandstone Formation times showing reconstructions of the reptiles in an imaginary scene with typical late Triassic plants."[7] The actual fossil specimens show few signs of predation, as they might have done had other creatures preyed on their remains. The authors have used the remains of an atypical mass death event to infer what the typical living situation might have been. As the authors remark, creatures of different ages appear to have died at the same time. This argues more for catastrophe than for the continuity of normal life. Had Murchison not rejected evolution out of hand, he would have had to think about the existence of these reptiles before their extinction and, even as a successionist, he could scarcely have believed that they only flourished at the time of their death. In other words, it is easier to think that they had been around for a long time than that they were entirely contained within a narrow band of Triassic rock. The smaller the area in which the reptiles were found the more necessary is an explanation of their having been isolated on it. If Murchison had bothered to think about this, he would at least have been able to ask the palaeontologists a few awkward questions. By this time, however, his research days were over.

---

[7]Ibid., 225.

The real question, though, is to determine *where* the catastrophe occurred that accounted for the sudden deaths of the representatives of several species. Not, certainly, at the longitude and latitude of the Moray Firth's present location, because two hundred and fifty million years ago the present continents and oceans had not yet been formed, and the land mass now called the British Isles was not only on or close to the equator, but was also joined to those other land masses in which the fossilized remains of relatives of the Elgin reptiles have been found. As usual in the history of geology, we have to propel ourselves back in time in order to imagine the circumstances that prevailed during the period that interests us, in this case the late Permian and early Triassic. This is hardly the place, perhaps, to review the geologic evidence that identifies the process of change in the configuration of land mass and ocean from, say, the Ordovician, through the Silurian and Devonian, to the Permian and Triassic, except to remind readers that the Old Red Sandstone which so interested Murchison had originally been deposited elsewhere on the globe to make what is sometimes called the "Old Red Continent," its fractures, faults, subsidences and suprapositions (to use Murchison's word) having occurred later, and also to recollect that both the building of mountains (such as the ones on which he worked in the North-West of Scotland) and, separately, the deposit of the so-called coal measures during the Carboniferous period had also happened later. A lot had occurred during the previous five hundred million years that Murchison did not know about. Cornelius Gillen writing about "The Geology and Landscape of Moray" in *Moray Province and People* says:

When these rocks (the Old Red Sandstone) were forming 360–400 m.y. ago, Scotland lay in the center of a large continent consisting of Britain, Scandinavia and North America. The latitude of the Orcadian basin was then around 20° south of the equator and the climate was semi-arid.[8]

In the novel *The Crow Road* Iain Banks gives imaginative expression to the same crucial idea:

Within the oceanic depths that lay beneath the surface of the present, there had been an age when, appropriately, an entire ocean had separated the rocks that would one day be called Scotland from the rocks that would one day be called England and Wales. That first union came half a billion years ago. Some of those rocks were ancient even then; two bil-

---

[8]Gillen, C., 19 , p.

lion years and counting, and shifting and moving across the face of the planet while that primeval ocean shrank and closed and all that would become the British Isles still lay south of the equator. Compressed and folded, the rocks that would be Scotland—by then part of the continent of Euramerica— held within their crumpled, tortuously layered cores the future shape of the land.[9]

In other words, long before achieving its present location, Scotland was situated in a place where the "Elgin reptiles" could flourish during the early Mesozoic. They had lived near the equator. Only as fossils had they traveled to Elgin.

The point we want to make can be put in another way. When a difference of opinion arose between Huxley and Murchison about the age of the rock in which the Elgin reptiles had been found, Huxley did not know any more than Murchison that they had been found in an a-typical Triassic outcrop. He had not inspected the terrain himself. He could not tell whether or not other fossil remains would be found within it. He only knew that similar fossils had been found in other parts of the world in Triassic rock. What Huxley actually said, in his letter to Gordon dated 11 November 1859, was:

I do not mean to say that what we know of Palaeontological Law at present warrants me in saying that the Elgin Reptiles *must* be Triassic—for none of them are known Liassic or Triassic genera.

Nor would I pretend for one moment that they *cannot* be Devonian—I, for one, see no *cannot* in the case. . . .

But the case seems to me to stand thus. It is granted that the stratigraphical evidence is not decisive either way. The Palaeontologist must therefore disregard it & be guided solely by the affinities of the fossils. Now the nearest allies of *Hyperodapedon* & *Stagonolepis* are admitted on all hands to be the Liassic or Triassic *Reptilia*.

Therefore until the contrary is proved, the presumption is that the Reptiliferous beds are Triassic at latest.[10]

At latest. As Huxley, like Benton and Walker, pushed geology aside in order to function primarily as palaeontologist, so Murchison, instead of beating the retreat that is represented by the fourth edition of *Siluria*, might have pushed the palaeontologists aside in order to reconsider his own field work, which despite his many visits had not resulted in his describing the terrain accurately. It is cruel to say this, because he *had* returned in 1858 and 1859 to engage in exactly

---

[9]Banks, I., 19 , p.
[10]DDW 71/859/43. *Huxley at Work,* 116–17.

that type of reconsideration. Let others say that he manipulated his results. The Murchison whom we know repeatedly reconsidered his research findings. That we can demonstrate that he did so—and this whole book is such a demonstration—does not remove the problem. If he went back to Moray time and time again, why did he not arrive at a fuller picture of the geologic complexity of the region? The only answer that one can fairly give is to say that, in order to learn from Murchison's field work, and to reconstruct it historically, one has to know exactly what he did or tried to do on these return visits.

Murchison has sometimes been criticized for his dependence on the cross-section, yet we have seen that he was far from being alone in regarding the cross-section as a useful tool. We have also seen that, quite reasonably in some respects, he depended not only on his early observations as recorded in his field notebooks, but also on early cross sections, until such time as other research, sometimes his own, sometimes other people's, proved them invalid. The cross-section is fine for purposes of demonstration, but is it all right as a research tool? We ask ourselves what the result would have been had Murchison walked in more or less a straight line from Covesea through Linksfield to Scaat Craig, or from Lossiemouth through Findrassie to Quarry Wood? Would it not have been logical to change the line of the cross-section, perhaps to one of the lines just mentioned, in order to test the original hypothesis? The directional line of a cross-section, if the strata are to be deduced from outcrops, must seem arbitrary, unless the purpose is clear and the reasons for the direction demonstrable. Murchison's purpose was to demonstrate the continuity of the Old Red. This was his chief concern. But when he returned to Spynie, Findrassie and Lossiemouth to seek evidence of two different types of sandstone meeting each other, he did not find that evidence. The simple rules of stratigraphy that he had learned at the beginning of his career argued for conformity of the reptiliferous sandstones and the Old Red Sandstone and as geologist he underestimated the complexity of the task confronting him which, however, is not to say that he was entirely mistaken.

When Murchison felt the need to check his data, it was nonetheless sensible to return first to those places where reptilian remains had been discovered because it ought to have been exactly at those places that different kinds of rock could be observed. Other people also did this and it may be useful to recollect what they said. For example, Harkness had said in his 1859 paper "Of the Yellow Sandstones of Elgin and Lossiemouth" delivered at the Aberdeen meeting of the British Association:

The rocks are to a considerable extent masked by debris; but whenever these are apparent, they manifest no traces of faults of such an extent as would disconnect the *Holoptychian* yielding strata from the reptilian beds which occur in this portion of Moray.[11]

After visiting Elgin in 1861 specifically for the purpose of addressing this problem, James Nicol wrote: "But at present I can see no stratigraphical reason for separating the reptile sandstones from the other sandstones of the district."[12]

In 1862, Joass told Gordon by letter that he and his friends on the Black Isle had vigorously retraced their steps in search of "Triassic rock" but eventually had reported to Murchison that his own original observations on that side of the Moray Firth had been accurate, prompting Murchison to say in a letter to Joass that their discoveries had "brought to demonstration by fossil evidence what I have always contended for that the yellow Reptilian sandstones of Elgin have their true equivalent in the yellow sandstones of Tarbat Ness and Dornoch."[13]

Earlier he had reported in Aberdeen that Professor Ramsay, during their visit to Elgin together, had endorsed Murchison's conviction that "no stratigraphical evidence exists by which the strata containing the reptiles can be separated from those charged with *Holoptychian* and other fossil fishes."[14]

While perhaps these quotations and references taken out of context have to be viewed with caution, they at least show that the opinions of Duff, Gordon and Murchison had not been undermined either by Joass and his companions on the Black Isle, or by well-qualified outsiders like Professors Nicol, Ramsay and Harkness. For them the nub of the question was:—Given that the reptilian remains had been found in only one type of sandstone, did that by itself indicate the existence of a Triassic stratum overlying the Upper Old Red?

Later in the century a different group of geologists made a determined attempt to find the point of contact between the two types of sandstone supposed to be one on top of the other at Cutties Hillock quarry, having a deep hole cut down below the level at which the quarry had been worked, but they, too, failed to find the evidence that would give the lie to Murchison's assertions about continuity and sequence. In other words, we are talking here about a real problem, not a supposed mistake of Murchison's.

---

[11]Harkness, 1859, p. 110.
[12]Nicol to Gordon 15 August 1861, DDW 71/861/9.
[13]Murchison to Joass 23 December 1862, DDW 71/862/54.
[14]Murchison's Aberdeen address. See Appendix 5.

To look again at Huxley's 1869 article on the *Hyperodapedon* is instructive at this point. The material he discusses in "On *Hyperodapedon*" had been at hand since early in 1867, *before* the fourth edition of *Siluria* appeared. But the talk was delivered and the paper published only in 1869—an unusual delay given Huxley's normal publishing habits. The article he had published in 1867 "On a New Specimen of *Telerpeton Elginense*" made no reference to the general stratigraphical questions that entered into his *Hyperodapedon* paper. Only when he turned to the *Hyperodapedon* did he allow himself those general observations that are so relevant to the present book. Huxley's paper is in three parts: (i) introductory remarks referring to the new evidence that gave rise to the paper; (ii) a description of "the most important remains of *Hyperodapedon*"; (iii) a few "general considerations."

In the course of his introductory remarks, Huxley itemizes the evidence of which he says he had recently become aware: first, several specimens sent by Gordon in 1866; second, two specimens from Warwickshire that Dr. Lloyd had lent him; and third, specimens from India which had been in the Geological Society since 1860 and to which Professor Oldham had drawn his attention only several years later. As to these Indian specimens he says:

The peculiar interest of this discovery arises not only from the sudden, enormous extension of the distributional area of *Hyperodapedon*, but still more from the circumstance that Dicynodonts have been found in the same Indian strata, and, thus, that we get a step nearer to the determination of the age of the remarkable reptiliferous formations of Southern Africa, the Triassic or Permian age of which was already highly probable.[15]

This same line of thought is picked up again in the "general considerations" which conclude the *Hyperodapedon* paper. In particular, four paragraphs seem to us so germane to the issue that we will quote them here in full.

The question of the terrestrial habit of the *Hyperodapedon* assumes a great importance when the wide distribution of the genus is taken into consideration. It has now been discovered in the North of Scotland, in the centre of England, and in central India; and if it were, as I doubt not it was, a terrestrial or semi-terrestrial animal, that alone indicates the existence of a very extended mass of dry land in the Northern hemisphere during the period in which it lived. And the proof of the existence of continental land in the Northern hemisphere acquires increased interest when we consider the evidence which shows what period this was.

[15]Huxley, T.H., 1869, p. 141.

The cardinal fact in that evidence is the occurrence of *Hyperodapedon* in the Coton-End Quarry in Warwickshire, as proved by Dr. Lloyd's specimen. It has never been doubted, I believe, that the Sandstone in which this quarry is excavated is of Triassic age. It has yielded Labyrinthodonts and Thecodont Saurians; and its stratigraphical position is such that the only question which can possibly arise is, whether it is Triassic or Permian.

As next in order of value, I take the discovery of *Hyperodapedon* in the Devonshire Sandstone, the determination of which as Trias rests, as Mr. Whittacker will inform you, upon independent grounds.

Thirdly comes the occurrence of the closely allied *Rhynchosaurus* in the Trias of Shropshire—a fact of subordinate value, but still by no means to be left out of sight.

These facts leave no possible doubt, as it seems to me, that *Hyperodapedon* is a reptile of Triassic age; but whether it is of exclusively Triassic age or not, and therefore competent to serve as a mark of the Triassic age of the deposit in which it occurs, is quite another matter, and one respecting which it behooves us to speak very cautiously.[16]

A few paragraphs later on he adds:

In discussing the age of the Elgin sandstones in years gone by reflections of this nature led me always to admit the possibility that these problematical beds might be of Devonian age; for *Hyperodapedon, Stagonolepis,* and *Telerpeton,* though clearly allied to Triassic and Mesozoic genera, were distinct from them, and have no stronger affinities with Mesozoic Reptilia, than the *Proterosauria* have (which yet are Palaeozoic and not Mesozoic), or than some of the Labyrinthodonts of the Coal (e.g. *Anthracosaurus*) have with those of the Trias.[17]

Given Huxley's knowledge of Murchison, this is extremely cool. Except when it provided the opportunity for doing battle with churchmen, Huxley in 1869 had not yet been fully converted to evolution. Not evolution but logic was the reason for Huxley's not closing his mind to the possibility of an earlier date for the Elgin reptiles. In these circumstances, it is difficult not to suppose that he took pleasure in emphasizing his own open-mindedness *after* Murchison had perpetrated his *volte-face* in *Siluria,* reflecting, as he probably did, that there had indeed been earlier discussions to which Murchison might have reacted more shrewdly. No indulgence in exclamation marks in Huxley's case. Just be careful about what you say. He knew that Murchison had not been. But Huxley found a means to steer around open conflict, before a knowing audience, in the deceptively bland remarks that are quoted above—"deceptively

---

[16]Ibid., 148.
[17]Ibid., 149.

bland" because he was much more outspoken in private. Characteristic is a letter of his to J.D. Hooker: "The great R.I.M. gave me his
address some time ago, but as you will not suspect me of having
read it, you will not wonder that I had not been aware of his iniq-
uities. I eat his salt yesterday at a lunch in honour of the Queen
of Holland, so I suppose I must not abuse him, but he is a very
trying old party."[18] (Back from the Himalayas, "brown, bearded
and brutal in health," Hooker had complained to Huxley about
Murchison's praising Hugh Falconer's work there—"far *beyond*
Thomson and myself"[19]) Ironically, it had been Murchison who,
through the good offices of George Gordon, had persuaded Grant
(schoolmaster at Lossiemouth) to lend Huxley the *Hyperodapedon*
specimens that allowed him to write the article referred to above.
Once again we are struck by the great difference between the con-
vention or, one could say, quite normal in-fighting of professional
metropolitan scientists and the more disinterested, if less focused,
practices of Gordon and his friends.

Where in our judgment Murchison went too far can be seen in his
letter to Gordon dated 24 January 1867 where he says: "I now see
that the enormous quantities of Lias fossils with which your region
abounds, had in these uppermost sandstones of the Trias or New
Red a natural foundation."[20]

He *saw* no such thing. Whatever he meant by the term "natural
foundation," he had certainly not observed it, and it is ultimately revealing that he was more willing to give way and lose face on the
question of the age of the reptiliferous sandstones (referring back
unglamorously and unconvincingly in his letter to Gordon to his
own earlier publications) than to abandon his belief in the generally
to be expected sequence of the principal strata, even though all the
evidence—to say nothing of Huxley's spoken speculations—now
pointed to the Moray Firth being unusual, not "natural," in a number of important respects. And so it has turned out to be. Consequently (though we cannot be sure) he may have read the list of
questions in Gordon's late January reply as no more than an innocent impromptu response, and not understood that he was being severely challenged, as a geologist, by another geologist who by no
means thought of himself as inferior, least of all in knowledge. Like
Huxley, Murchison had taken Gordon too much for granted, prof-
iting from his disinterested help but perhaps not fully appreciating

[18]ALS Huxley to J. D. Hooker, 19 July 1865; Kew Library.
[19]ALS J. D. Hooker to Huxley, 10 October 1865; Kew Library.
[20]Murchison to Gordon 24 January 1867, DDW 71/867/3.

that he had things to say on his own account. Yet in the final analysis, Gordon may not have been bothered by this. He continued to study the geology of Moray for a further twenty-five years or so in the company of experienced experts like Geikie, Horne, Benjamin Peach, Judd and Newton. It seems a great shame that Murchison, having enjoyed a comradely relationship with Gordon, and having been so energetic around the Moray Firth for four decades, should have bowed out in such a sad way.

The letters reproduced at the end of Part Two express the shock experienced by Murchison's old friends in Moray when they heard about his capitulation and subsequent revision of *Siluria*. "The shock was all the more severe," wrote Joass to Gordon, "coming as it did during the pleasant excitement of the *Telerpeton* triumph. It made my heart sore to think of Sir Roderick's defection from the Field and his alliance with slippered Palaeontology."[21] Defection from the field; a harsh comment on either an army officer or a geologist. Gordon shared this sense of outrage and seems to have dispersed many copies of Murchison's final letters to him. We ourselves, however, recollect how Murchison began his career. Almost like a tragic protagonist, he showed awareness of the weakness that was to bring about his downfall. As he started to write *The Silurian System* he despatched an interesting letter to Professor Webster. "I had not got further than the second chapter when I found it was so very *palaeontological* that it would be *essential* for me to have as my consulting editor someone acquainted with all the references and synonyms of organic remains."[22] Given the resources of the School of Mines, the Geological Society and the British Museum, to mention only English institutions, his failure during the next thirty-five years to strengthen his understanding of palaeontology, at the same time only slightly modifying and improving his practice in the field, must remain a disappointment. Some colleagues who noticed his limitations, having limitations of their own, preferred to get on with him in a professional and comradely fashion as best they could. Huxley, however, lacked such delicacy. He saw Murchison's weak point and exploited it.

---

[21]ALS. Joass to Gordon, 28 January 1867, DDW 71/867/5.
[22]ALS. Murchison to Webster, 6 December 1842. APS. Although the American Philosophical Society gives this letter an 1842 date, it is almost certainly 1832 and must be before 1839 when *The Silurian System* appeared. Murchison offered Webster an inducement of 10 pounds.

7. CONCLUSION

We intend that the two parts of this book will be indispensable to each other, or rather that the two parts will be indispensable to the reader, the discussion and the documents together defining Murchison's life and work. We believe that scientific documents should, if possible, be allowed to speak for themselves, especially when they are being published for the first time. In Part Two we therefore reproduce the Gordon-Murchison correspondence in its entirety. We do not, on the other hand, believe that there is only one way to create the context in which such documents can best be understood. Possibly there are several ways, or modes of interpretation. In this book we have tried one, but not in a doctrinaire manner.

From where we have chosen to locate ourselves, we can observe much that is of interest. First, we see Murchison living through that phase of the professional development of geology in Britain that is associated with the establishment and administration of the Geological Survey of Great Britain. We have noted that the Geological Survey of Scotland, which he did so much to promote, did not reach North-East Scotland until after his death. Although Murchison fostered, built up, argued for, staffed and developed the Survey during his period as Director, helping to give it the prestige it needed in order to flourish, as well as significantly increasing the size of its personnel, he himself seems never to have refined his technique in the field as his staff was obliged to do, depending instead on what he had learned with such great enthusiasm from his early mentors. His research in Moray was interrupted by other activities; five years on Russia and then fifteen years as President of the Royal Geographical Society. If the North-East of Scotland was one of his laboratories, he was away from it for too long, and by the time he returned he found it difficult to adjust to the scientific advances that had been made in his absence. Perhaps in his private letters to George Gordon his dilemmas are more clearly exposed than in his official correspondence.

Second, he was not deeply religious, having the same doubts about Christianity as many other Victorians, whose observance of its rituals was sometimes merely superficial. Murchison denied himself full intellectual freedom by remaining a so-called Successionist, thus anchoring himself to the period in which he did his main work. Everything had been created by God but not at the same time. Compromised by this inadequate and chronically superficial theory, he found it difficult to think about many of the problems that confronted him, such as that of the origin and extinction of species. The letters that follow show that his mind was closed to evolution. Already sixty-seven when the *Origin of Species* appeared, its implications seem not to have interested him. He was nostalgic for the old order and not excited by the prospect of the type of social change that would have been consistent with new knowledge. This made him seem old, curmudgeonly and unsophisticated in the eyes of his younger colleagues, and perhaps in Gordon's eyes as well. The letters show the difficulties a person will often have if he or she lacks a well thought-out position on matters of immediate intellectual importance.

Third, from our particular vantage point, we believe we can observe the effects of the interplay of personality on the way in which the geological problems of Moray were tackled. The texts produced by Murchison and Huxley, and to a lesser extent by Gordon and others, in what amounted to an aggressively conducted, albeit ostensibly polite debate about the age of the Elgin sandstones was simply the most obvious element in this clash of minds and wills. The expression of ideas was also the revelation of character; the more urgent the debate the more characteristic were the individuals' contributions to it. Someone might argue that the participants in the debate were not free spirits but represented socially or politically identifiable positions—with Murchison unable to think about instability and extinction because of his adherence to an immutable class system, and Huxley inhibited by his belief in logic and in the subversive as well as the creative importance, socially, of science. Thus what seemed to be a straightforward argument about the interpretation of geological and palaeontological evidence was in fact ideologically complex and interesting. We would nevertheless hold that these previously unpublished letters throw fresh light on the motivation and entanglements of the participants. Murchison was unnecessarily stubborn. Huxley was unnecessarily aggressive. Gordon was unnecessarily reserved. Good-hearted collaboration would have produced better results, and more quickly. Of course geologists, like the rest of us, have to decide, if they can, whether to keep their minds in motion or not. If social pressures and determinants exist, the mobile mind may take stock of them. The snag is that this forward movement will be arrested, albeit temporarily, when words are written on a page and texts are published. A deep compulsion made Murchison protect his texts, thus reducing his ability to respond to other people's ideas, though respond he did, after a fashion.

Finally, we note that, in this case at least, geological understanding was advanced both by metropolitan specialists for whom the publication of results was supremely important and by people of intellect in other parts of Britain who, though living well away from the main centers of academic activity, nonetheless made valuable contributions to the subject by acquiring a keen perception of its problems and by engaging in useful, often extremely energetic field-work. Gordon's personal contribution to geology in Scotland is to be treated in a biography. Although content to spend most of his time in the North, he retained many of his Edinburgh and Glasgow contacts throughout his life, traveling farther south to the museums and libraries of England and Ireland when he felt the need. The evidence suggests that he had local access to both Scottish and English journals, his knowledge of what was happening in the south being greatly facilitated by the postal service and the growth northwards of the railway system. He is an example of the best kind of intelligent, erudite and well-educated Scot—one of many others including his own friends and associates. Gordon, as much as Murchison, is the link between Part One and Part Two of the book, insofar as his correspondence was a research instrument whose dialectical character helped significantly in the identification of geological problems and the assessment of possible solutions.

PART II

A NOTE ON TRANSCRIPTION OF THE LETTERS:

1. All the letter-texts are reproduced verbatim and in their entirety. Despite the inconsistencies and irregularities of the originals, the intention has been to keep editorial interference to a minimum.

2. The punctuation has not been standardized, though a dash has occasionally been rendered as a comma or a period when a twentieth century reader would expect one or the other.

3. Square brackets have been used to indicate the presence of words, phrases or sentences that have proved indecipherable. We have preferred to indicate severe difficulties of transcription in this way rather than resort to guesswork.

4. In a few instances, square brackets are also used to indicate the insertion of words which the sense requires but which the author omitted.

5. Round brackets and question marks in any letter text were in the original autograph. We have not used the question marks editorially to indicate doubtful readings.

6. The editorial device "sic" is not used. Where there is a discrepancy between the spelling of a name in a letter text and the spelling of the same name in a note, the note is correct.

7. The library reference number is placed at the head of each letter. In the case of Elgin letters (those beginning with DDW 71) whose place in the sequence should, in our judgment, be different from that determined by those who originally catalogued the collection, we have retained the Elgin number and explained the shift in sequence in a note. Sometimes the same reference has been given to two or more letters, either because they are on the one piece of paper, or because of a cataloguing error. Here, too, we have retained the Elgin number. Readers should note that Michael Collie and Susan Bennett are currently making a new catalogue of Gordon's correspondence for publication by Scolar Press in 1995 and that the old DDW numbers are being replaced in order to permit an accurate chronological listing which incorporates a substantial amount of previously uncatalogued material. In the new catalogue reference will be to the year and the chronological position of any particular letter within that year. Meanwhile in the present book the reference numbers beginning LGS/M refer to the Murchison correspondence in the Library of the Geological Society of London. The reference numbers beginning ALS refer to the Lyell correspondence in the American Philosophical Society archives.

8. The dates given at the head of each letter have been taken directly from the autograph manuscript, except in the cases noted, and have been standardized as to order, i.e. day first, then month, then year.

9. To save pointless repetition, the address at the head of each letter has been omitted. All of Gordon's letters were from the Manse at Birnie. Most of Murchison's were dispatched from Jermyn Street. Whenever this was not the case we have noted the fact.

10. We have not transcribed the signatures, since they are uniform throughout. Murchison always signed his letters "Rod I. Murchison". Gordon always signed his letters, "George Gordon".

11. We record deletions from Gordon's autograph drafts only when we believe them to be significant or interesting. Many such deletions have been rendered indecipherable to the naked eye.

12. Gordon's notations on letters received from Murchison have been noted when germane to the narrative.

ABBREVIATIONS USED IN THE BIBLIOGRAPHIES

Annals of Science	Ann. Sci.
Annals of Scottish Natural History	ASNH
Archives of Natural History	ANH
British Journal for the History of Science	BJHS
Edinburgh New Philosophical Journal	Edin. New Phil. Jrnl.
Geological Magazine	Geol. Mag.
Journal of the Geological Society of London	JGSL
Journal of the Society for the Bibliography of Natural History	JSBNH
Memoirs of the Wernerian Society, Edinburgh	Mem. Wern. Soc. Edin.
Philosophical Transactions Royal Society London (B)	Philosophical Transactions
Proceedings of the Geological Society of London	PGSL
Proceedings of the Geologists' Association	PGA
Proceedings of the Royal Physical Society of Edinburgh	Proc. Roy. Phys. Soc. Edin.
Proceedings Zoological Society of London	PZSL
Quarterly Journal of the Geological Society of London	QJGS
Reports of the British Association for the Advancement of Science	BA
Scottish Journal of Geology	Scott. Jl. Geol.
Transactions Edinburgh Geological Society	Trans. Edin. Geol. Soc.
Transactions Geological Society of London	Trans. Geol. Soc. Lond.
Transactions Highland and Agricultural Society of Scotland	Trans. High. Agr. Soc.
Transactions Inverness Scientific Society and Field Club	Trans. Inv. Sc. Soc. & Fld. Club
Transactions Royal Society Edinburgh	Trans. Roy. Soc. Edin.

THE LETTERS

To: Malcolmson, Forres
From: Murchison, 2 Eccleston Street
Dated: 4 January 1839
DDW 71/839/1

I am much gratified with your letter of the 12 Decr, & think that your "pieces petrificatives" will (from your statement) entirely bear you out in your inference, that the Red Sandstones & schists of Cromarty, Forres, [Spey Bank] & [Gammrie] are parts & parcels of the Old Red System. To make your work & that of Mr Gordon complete, you should run your enquiries <u>fairly up to Gammrie</u>, & if you do so you should endeavour to show <u>how</u> it is, that the nodules with fish occur there in horizontal strata on the inclined beds of the Old Red Sandstone as represented by Prestwich.

My speculation is that the said fish beds occurring in a ravine may prove to be nothing more than a regenerated or ancient alluvial deposit made up of Old Red Detritus. In short, that spot requires close examination in connection with your Spey discoveries, & then your problem will be completely demonstrated.

I beseech you to stick to the work & to prepare a memoir either by yourself or in conjunction with Mr Gordon for the Geological Society which when you come to Town in spring you can [brush] up by zoological & other comparisons. No one has hitherto really explained the Murray & Elgin tract. Sedgwick & self merely glanced at it, for the purpose of <u>general comparison</u>; for we had a great duty to perfom in subverting a most erroneous postulate of MacCullochs, viz that a vast portion of the Red Sandstones of the Highlands are "<u>Primary Sandstone</u>." Caithness gave us the key & the fossils proved, that the whole mass was nothing more than the equivalent of our English Old Red.

In such an enquiry you will doubtless have many new organic things, all of which will be done justice to if you bring them to London. They will be drawn & [   ] in the Geol. Transactions doubtlessly.

My great work is now before the world & you can see it when you come to Town. The only Northern Subscriber I have on my list is Sir George MacKenzie. If authors were rewarded this work ought to have been charged £12. In some future edition I may make use of your present labours. Every discovery, however, which extends the [grandeur] & confirms the individuality of the Old Red Sandstone is of high interest to me. Foreigners have no notion of this vast British System.

Let no one rob Mr Gordon & yourself of your desserts, so get up your memoir forthwith & <u>condense</u> as much as possible.

To: Gordon
From: Rosehaugh, Munlochy
Dated: 5 September 1858
DDW 71/858/20

I beg to enclose a letter from our mutual friend D^r H Falconer[1] which explains the cause of my call at Elgin & its environs.

I hope to be at Elgin on Tuesday morning at farthest & will enquire at the P. office for any note you may address to me.

I [shall] but little time to devote to the habitat of the *Telerpeton*[2] but must endeavour to satisfy myself that this animal is really of 'Old Red' age.

To: Murchison
From: Birnie
Dated: 14 September 1858
DDW 71/858/21

I have just packed up a box containing (together with the pocket compass which you lost at Covesea) the following specimens and casts. I beg your acceptance of them.

---

[1]Hugh Falconer (1791–1867), palaeontologist and botanist, was superintendent of the Botanic Garden at Saharanpur, India (1832–42) and Professor of Botany at the Calcutta Medical College (1848–55). FRS 1845. According to *DNB*, Falconer retired in 1855 and traveled in Europe before settling down in Britain.

[2]This apparently innocent remark reveals Murchison's state of mind in 1858. G.A. Mantell's "Description of the *Telerpeton Elginense*, a Fossil Reptile Recently Discovered in the Old Red Sandstone of Moray" (1852), and other articles produced in the early eighteen fifties, had alerted Murchison to the possibility that his earlier research might require revision, but at this stage he did not believe he could be seriously mistaken. Huxley was already at work on *Stagonolepis*, the significance of which Murchison did not yet appreciate, convinced as he was that, if his stratigraphy was accurate, he could afford to ignore the palaeontologists. This is why in letters to others he disparagingly referred to Huxley's reptiles as "the frogs."

(2) Mr Young[3] No 1 to your Museum. (1) The two marked N° 10 and N° 11, being duplicates, of parts of the large slab (N° 6) I should wish to make offer of to Prof^r Owen[4] if you think them worth his acceptance.

N° 1   Footprints on the Sandstone of Cummingston Quarry near Burghead, Elginshire N.B.
          Presented to the Museum Jermyn St London by Alex^r Young Esq Fleurs by Elgin.
N° 2   Block of Lossiemouth Sandstone, with bone and <u>large plate or scale.</u>
No 3   Part of plate or scale, Smiths quarry Lossiemouth.
N° 4   Four small specimen from d^o.
N° 5   Cast of the specimen (now in the Elgin Museum) which Agassiz named *Stagonolepis Robertsoni.*
N° 6   Cast of Slab, now in Elgin Museum, discovered by Alex^r Young Esq at Findrassie.
No 7   Part of ditto position marked by x.
N° 8   Cast of a scale in the possession of Alex^r Young Esq.
—9   Cast of
N° 10}   a bone
      } Casts of and for Prof. Owen
   11}   scale

To: Gordon
From: Jermyn Street
Dated: 8 October 1858
DDW71/858/23

I have to thank you abundantly for the transmission of the boxes from Elgin.

There is no doubt in the minds of both Owen & Huxley that the bones are those of Reptiles—the existence of which is so clearly marked by the footprints.

---

[3]Alexander Young of Fleurs, Elgin. With reference to the letter of 20 October (following), it is important to distinguish here between specimens and casts. The casts were presumably made of plaster and were difficult to use in London. Enough original material was, however, sent in this consignment for the Jermyn Street imagination to be kindled. Huxley had probably not seen whatever specimens Murchison, Malcolmson and others had taken to Paris in the eighteen thirties; Louis Agassiz, when he classified the *Stagonolepis* as a fish, had not seen the original fossil, but only a drawing of it. Thus the story begins anew in 1858 when more specimens are found and more people see them.

[4]Sir Richard Owen (1804–1892), was an English zoologist and Hunterian Professor of comparative anatomy at the Royal College of Surgeons. In 1858 he was at his professional peak as President of the British Association at Leeds, Fullerian Professor of Physiology at the Royal Institution, lecturer at the School of Mines, and Superintendent of the Natural History collections at the British Museum. He had fundamental differences with Darwin and Huxley concerning the origin of species, preferring the notion of continuous creation by a deity.

I have only one regret which is that in your anxiety to let me have the things quickly the plaster casts were packed too soon & that their moisture affected everything, so that Peachs[5] compass was just as if it had been fished up off Lossie Mouth & your nice labels were much injured though all capable of being read—The casts (I regret to say) of the great slab with the most important of all the bones is broken, but we are getting it put together.

You shall hear again when the Anatomists[6] begin to speak out.

I wish it were possible to get a really good flagstone with the clear markings of the large (Shetland Pony)[7] animal as well as of the smaller.

The footprints on the sandstone you sent cannot have pertained to an animal larger than a puppy whilst the bones announce a huge creature.

The casts lent to me by M[r] Patrick Duff[8] have nothing to do with bony or vertebrate animals—They belong undoubtedly to great Cephalopods— (Orthoceratites?)

The enclosed paper is an abstract of what I did & said in Geological matters relating to the North of Scotland at Leeds.[9]

Details will be given hereafter to the Geological Society with Sections etc.[10]

To: Gordon
From: 16 Belgrave Square
Dated: 10 October 1858
DDW 71/858/24

We must not be beaten by M[r] Beccles[11] & in enclosing you his letter I request you to spend £5 or £10[12] or more if necessary in extracting & forwarding in cases to me in Jermyn St slabs exhibiting clear & good footprints of various sizes.

---

[5]Charles Peach (1800–1886) coast guard and comptroller of customs at Wick.

[6]i.e. Huxley and Owen.

[7]Presumably, this is an early reference to the tracks later attributed to *Stagonolepis*, see Huxley's paper "On the Stagonolepis Robertsoni (Agassiz) of the Elgin Sandstones; and on the Recently Discovered Footprints in the Sandstones of Cummingstone," *QJGS*, 15, (1859): 440–60.

[8]See Part 1, page 69.

[9]Murchison's talk at the British Association meeting in Leeds in 1858.

[10]See Appendix I, Items 19 and 20.

[11]Samuel Husband Beccles FRS (1814–91). At the same 15 December 1858 meeting of the Geological Society to which Murchison presented a paper on the sandstones of Moray and Huxley on *Stagonolepis*, S.H. Beccles gave a talk "On Fossil Footprints in the Sandstones of Cummingstone" an abstract of which was published in *QJGS* Volume 15: 461. He reported that he had sent "a large collection" of footprint slabs to London, but had not at that time had the opportunity to study them in detail. Murchison's response was somewhat less than chivalrous.

[12]Murchison usually paid amounts of this kind from the funds of the Geological Survey, as he probably had in mind to do on this occasion, but at other times he may have paid Gordon small sums from his own pocket for expenses incurred locally.

I wish to have you & your coadjutors out in the foreground & Mr Beccles will see by my printed notice[13] that we are all before him.

Still it is highly desireable to have some of these slabs exhibiting different species in my Museum—i e in the Public Museum of the Gov^t. School of Mines.

There is I hope yet time to get them before the bad weather sets in. I enclose a separate paper for Mr A. Young who will I am sure be as obliging to me as he or some other proprietor was to Mr Beccles.

> Yours truly
> Rod I Murchison

Pray give a Notice to Mr Martin[14]

Mr Beccles is the Gentleman who first finding footprints of huge Saurians in the Hastings Sand (Wealden), has since been exhuming loads of little Mammals in the Purbeck Strata (uppermost beds of the Oolite).

What do I owe you already?

To: Gordon
From: Jermyn Street
Dated: 12 October 1858
DDW 71/858/25

I have just had our good friend Dr Falconer here, & he says that he has no doubt that on my representing to you that it will be of the greatest advantage to Science, that we should have here under our inspection for a month or so, the original slab of stone with the impressions of bones & dermal plates or scales that Mr Patrick Duff yourself & others will after holding a Sitting comply with the urgent request of the Director General.

In fact you will have the relics cleaned out by a first rate comparative anatomist Professor Huxley who is now engaged in studying all the other relics I have brought up or which you have sent.

Now the casts do not convey to the real naturalist the perfect idea of the whole truth at which he aims. If you find your associates & the Directors of the Elgin Museum ready to comply with my prayer I would beg that the Specimens be tightly packed in [lard] not saw dust with a good envelope of strong paper.

I am very very anxious on this point, as I wish the data to be clear when my new Edn of Siluria appears.[15]

If there is no packet sailing we do not mind expense & they should be sent at once by the Goods Train of the Rail Road to this place.

> Yours sincerely,
> Rod I Murchison

---

[13]See notes 9 and 10 to the letter of 8 October 1858.

[14]John Martin was known for his essay on the geology of Moray which was published in Volume V of the *Prize Essays and Transactions* of the Highland and Agricultural Society of Scotland in 1837. This same John Martin was the first curator of the Elgin Museum.

[15]The 3rd edition of *Siluria* appeared in 1859.

Falconer desires me to say that although he has not written he is looking after your interests.

(When the slab returns it will have proper names attached to all its parts.)

To: Murchison
From: Birnie
Dated: 14 October 1858
DDW 71/858/26

Inter alia "The presumed vertebral column turning out a Cephalopod gives our locality an additional interest.—In reading your printed notice, it occurs to me to state that as yet no footprints have been met with at the Covesea quarry. They have appeared only in that of Cummingston which belongs to our county member, and not to Mr A. Young who is a younger brother of the laird of <u>Burghead</u>."

To: Murchison
From: Birnie
Dated: 20 October 1858
DDW 71/858/27

Yesterday I packed up and dispatched three boxes for you by train to Aberdeen—thence by Steam boat. They will reach you on Saturday.

Box N° 1 contains the slab which you saw[16] in the Elgin Museum and which was found by Mr Alec Young at the <u>Findrassie</u> quarry. You were not at this place but were within ¼ mile of it when at the quarry (on our return from Covesea) when you first discovered that you had lost Mr Peach's compass.

The original specimen of *Stagonolepis* from Lossiemouth is also in this box. From this or a photograph of it M. Agassiz gave the name—These two belong to the Museum. A small tin box with a mineral which Mr Duff is most anxious to have submitted to some of your friends who will be able to say if it be organic or not.

Box N° 2 contains some slabs with impressions from the Cummingston Quarry, Burghead. They belong to Mr Alec Young of Fleurs, who kindly

---

[16]As a result of Murchison's visit, Gordon now sends to London the original specimens as opposed to the casts or impressions referred to in his letter of 14 September. Although Gordon from the outset makes a clear distinction between gifts and specimens on loan, many decades were to pass before some of the latter were returned to the Elgin Museum. In a letter to Gordon dated 4 January 1858, Huxley had recommended that Findrassie (a place he did not know and could not have located on a map) should be their "great hunting ground." The specimen from Findrassie mentioned in this letter is the local response to this request. This material, and the footprints which soon followed, allowed Huxley to write the important paper on the *Stagonolepis* referred to in note 7 to the letter dated 8 October 1858. The division of labour between Huxley and Murchison, representing crucial differences of approach, is discussed in Part One.

allowed me to send them to London for [inspection]. The contents of these two boxes are expected to return to the North greatly enhanced by their trip to London.

Box N⁰ 3 contains various specimens with casts of bones, plates etc collected by Dʳ [Jas] Taylor[17] H.E.I.C. & myself last Friday week among the debris of the quarry at Findrassie. I have Dʳ Taylor's concurrence in making offer of the whole for your acceptance—Some of the specimens will have to be broken up before they reveal all that I think is in them. The few slabs with foot prints (three pieces) were got by me at Cummingston on Saturday. I readily make them over to you.

To: Murchison
From: Birnie
Dated: [25] October 1858
DDW 71/858/28

After three visits, including two days of hard labour at the quarry, I am delighted to say that I have been able to secure and pack up six cases of footprints, which however owing to some irregularity in the London steam boats, will not reach you until this day week.

Some of these contain the finest, largest and most distinct impressions that Mʳ Anderson[18] or any of his quarrymen ever saw, and I consider myself very lucky in stumbling upon them.

The proper name of the locality is Mason's Haugh quarry, between Burghead and Cummingston close by the sea presently rented by Mʳ Anderson and on the property of Chas. L. Cumming Bruce[19] of Roseile Esq M.P. It is the only place where footprints have been as yet observed.

I was accompanied by Mʳ Alec Young who left no stone unturned to get the best for you. Mʳ Anderson the tackman of the quarry was most obliging, gave us every facility, sent his men to their work and [occasionally] gave us a helping hand himself.

I paid the workmen and when I get the carpenters account will let you know the amounts of cost.

The specimens in cases N⁰ 1, 2, the four pieces wrapt in paper in N⁰ 3, and the five large slabs in the triangular case N⁰ 4, are all from the surface

---

[17]Dr. Taylor of the Honorable East India Company. In all likelihood this was the same Dr. Taylor as accompanied Joseph Hooker on his travels through Northern India and Tibet. In a letter to William Hooker from Falconer, who was assisting J.D. Hooker while in India, there is a reference to Taylor's "beautiful drawing of Kanchum Junga. That was in May 1850." Gordon was personally acquainted with both Hugh Falconer and Joseph Hooker.

[18]Of course this is not the same person as Dr. John Anderson mentioned later in the correspondence.

[19]Charles Lennox Cumming-Bruce 1790–1875. His father was Sir Alexander P. Cumming-Gordon. On the death of his father, his elder brother changed his name to Gordon-Cumming. But when Charles married Mary Bruce in 1820 he changed his name to Cumming-Bruce, and was seven times elected MP for Morayshire.

of one bed. I hope you will be able by the black lines[20] put on them and their natural fractures, to put the whole of them in nearly their original contiguous position.

Cases 5 & 6 contain fragments taken from the same [corner] of the [quarry] but their position relative to the others cannot be ascertained.

I would suggest that the surfaces of the slabs should be washed and scrubbed to remove the clay dust etc. If examined under the light of a lamp or candle held in a certain position, the contour of the footprints come out very strikingly.

If you favoured me with a proof of your note for Siluria regarding the geology of this district, I would be able to correct, if necessary, the paper.

I have taken the liberty of sending you a copy of a paper drawn up by Mr Duff and to be read at our Society regarding the Reptilian remains of the Old Red Sandstone[21]. This paper was brought to me by Dr Taylor who thought you might be interested in what it records.

You are no doubt aware how anxious we all are to hear what is made of these animals by those of your friends able to decide—Let Falconer see them.

To: Gordon
From: [Jermyn St]
Dated: 29 October [1858]
DDW 71/858/30

Intending to present to the Geological Society a regular Memoir on my additional acquaintance with the Old Red of the North & particularly of the Morayshire deposits I shall in the sequel endeavour to do full justice to Yourself & all concerned whilst Prof Huxley[22] will describe the Organic Remains not only generally in my Memoir for the Geological Socy but in Special Decades as part of the Geological Survey.—i e with beautiful large plates & full Natural History explanations.

We are overjoyed at the prospect of the arrival of the 6 cases you have secured for us. They will cut a fine figure in our halls.

---

[20]Huxley's letter of 18 November 1858 acknowledging the receipt of six boxes from Elgin appears to be the answer to this. Huxley said: "You have done so much for us that it is a shame to criticize any of your arrangements but may I suggest that letter & numbers would facilitate putting fragments of slabs together much better than any other marks." Examples of these numbers and letters can be seen on several of the specimens in the Elgin Museum.

[21]Duff, P., 10 December 1858. Reptiles in the sandstone of Moray, *The Elgin Courant*, 25, 1252, p.6.

[22]As Gordon began to send, not casts, but the fossils themselves to Jermyn Street, Murchison was obliged to decide what to do with them. By the time he wrote this letter, he had delegated all the palaeontological research to Huxley, which is why, even after Murchison's death, Huxley continued to talk about having been "instructed" to write a Memoir for the Geological Survey.

In the meantime such is Huxley's passion to have the true originals that I have urged Mr P. Duff by letter of this day to let us have in our trusty keeping the *Telerpeton*[23] or *Leptopleuron* for a brief time.

I must also tell you that Huxley has made a capital cast out of one of the [hollows] & has produced a biconcave vertebra of such a form as induces him to consider the reptiles to be of high organization—possibly higher than Labyrinthodonts! Of all this anon.

He is most anxious for more <u>bones</u> to pick. If by employing a few pounds in excavating along the loose rubbly deep yellow beds at the <u>base</u> of the Lossiemouth cliffs it were possible to find a <u>tooth</u> or bit of the <u>skull</u>[24] of the animal whose fragments we picked up such remains would settle the whole question.

I enclose for your perusal & correction a little addendum to my notice in the Appendix to my forthcoming Siluria which if you can read, I wish you to correct.[25] I am overladen with work—Falconer is off tomorrow m[orning.]

To: Murchison
From: Birnie
Dated: 2 November 1858
DDW 71/858/31

I have the pleasure of returning your M.S. note with a few corrections. Your letter of 29th ult—with that to Mr Duff and Prof Huxley's one of the same date, will indeed induce me to be on the outlook for some of those pieces or casts of the skeleton which he Prof H so much desiderates. I shall visit Lossiemouth in the course of a day or two but purpose first to engage some men to turn over the debris of the old quarry at Findrassie in that part of it where Dr Taylor & I found those now with you—I paid this locality a hurried visit one day last week and got two slabs with impressions. I have strong hopes that some parts of the head will be got there.

---

[23]It was not until 1866 that Huxley had enough fresh information about the *Telerpeton* to write "On a New Specimen of *Telerpeton Elginense*", *QJGS,* 23, (1867): 77–84.

[24]In his letter to Gordon dated 16 November 1858 (DDW71/858/35) Huxley wrote: "The jaw with teeth is a wonderful thing, but very tantalising for as it is only a cast I cannot tell what was the skeleton of the teeth—(a great point in my enquiry) nor can I be sure of their mode of implantation which is no less important." This refers not to *Telerpeton* but to *Stagonolepis* which Huxley would soon be claiming was crocodilian. On 9 December 1858 he pursued this point in another letter exclaiming: "A tooth! A tooth! my kingdom for a tooth." Huxley made a habit of quoting Shakespeare out of context. How prophetic it would have been had he said "horse" but fourteen years were to elapse before he came to appreciate the significance of the evolution of the horse.

[25]Murchison is probably referring to Appendix B of the 1859 Third Edition of *Siluria* which is entitled "B.-Geology of the North of Scotland." Subsequent appendices, "Q-The *Stagonolepsis Robertsoni* of the Elgin Sandstone" and "S-Additional Notice on the Old Red Sandstone of Morayshire" are dated 27 November 1858 and 10 December 1858, respectively. In Appendix B, Murchison provides a brief history of the development of his thinking concerning the stratigraphic position of the Elgin sandstones. He relates the events surrounding the discovery of *Telerpeton* and *Stagonolepis* and footprints.

The large slabs etc footprints, from "Mason's Haugh" quarry near Cummingston should be with you today. It will gratify me much to hear that they have got to Jermyn St in tolerable condition, and to learn your <u>first impressions</u> of them, particularly of the largest one.

I can see how you mean to dispose of the strata. In the mode meted out for your arrangement of our strata I have some difficulty in disposing of the Inverugie concretionary rock (limestone). It certainly dips to the North.[26]

To: Gordon
From: Jermyn Street
Dated: 5 November 1858
DDW 71/858/32

It would be well perhaps that M[r] Patrick Duff should send his *Telerpeton* by rail—for we care not what the transit may cost & Huxley is most anxious to examine it.

The big slabs have not yet arrived; but they will be in all good time.

My Memoir will come off on the 1 Dec[r] followed by Huxley's description of the animals. He is very anxious, by the bye, to know if it be certain that the piece of yellowish or brownish yellow sandstone which contains the <u>hollow</u> or <u>mould</u> of a Vertebra is in the same rock as that in which scales of *Stagonolepis* were found.[27]

This specimen which is perhaps the most important yet sent (& of which Huxley has prepared a metallic cast) brings out a Reptile apparently of such very high organization that he H is alarmed; the more so as he hears of Jurassic & Wealden things near Elgin.[28]

State the [find] the locality & the relations.

To: Murchison
From: Birnie
Dated: 9 November 1858
DDW 71/858/33

On Wednesday I employed some labourers in turning over part of the debris of the Findrassie quarry, and was successful to the extent of picking up some twenty or more bits of flags with impressions of organisms, some

---

[26]This entire paragraph was deleted by Gordon as he was composing the letter.
[27]This would seem to show that Huxley, though at such a distance from the Moray Firth, was learning to distinguish between the types of sandstone to be found in the Elgin region, though his interest extended only so far as being able to place the fossils correctly.
[28]Gordon (1832) wrote about a large fossiliferous block composed of alternating marl and limestone at Linksfield. Malcolmson (1838) and Duff (1842) assigned it to the Lower Purbeck Beds of the Wealden. Additional discoveries led to papers by Moore (1860) and Judd (1873) in which glacial erratics containing Mesozoic fossils were described.

of which I am persuaded will throw additional light on the interesting enquiry in which Prof. Huxley is engaged.

In the after part of the day I was joined by Mr. Macdonald of the Elgin
Academy who detected the specimen with the cast of jaw and large teeth.

One at least of the other specimens shows also casts of teeth. Some will
require to be broken up. And one or two perhaps have no impression on
or in them. I made a search in vain to discover the bed in the Findrassie
quarry from which these slabs have been raised. The mounds of debris
must have been thrown out upwards of 40 years ago.

I trust you or Prof. Huxley will kindly drop me a note on their arrival. I
regret to hear that the large slabs from Mason's Haugh quarry, Cummingston, had not reached you when you wrote on the 5th. I trust you will
have got them by this time, if not let me know and I shall set about in-
quiring after them.

I read your note to Mr. Duff and both he and I are not certain to which
specimen you allude as containing the mould of a vertebra. If it has been
among the specimens I sent you it must have been from Findrassie. Or, if
it be the specimen you got from Mr. Duff and which he had sawn up it
was also from the same locality. We are not aware of any specimen having been given or sent to you except those from Cummingston, Findrassie
and Lossiemouth. Findrassie quarry, being on the northern slope of the
Quarrywood hill (or Bishopmill & Alves ridge) belongs to the inland and
not the coast escarpment. As you are aware we have so much drift etc.
that the relative position of our outcropping strata have to be guessed at;
but so far as I can judge the bed in which the *Telerpeton* were found and
the Findrassie quarry must be very near to each other if not the same bed.

I do hope that what I have now sent will remove all doubt of the bones
etc. belonging to the same animals as the Findrassie impressions of scales
and dermal plate which so closely resembles the scales etc. from Lossiemouth.

I am more then ever anxious to hear the result of Prof. Huxley's examination of the present parcel—and if any the least doubts remain on his
mind I shall be happy to turn out the other day with double the number
of labourers.

To: Gordon
From: 16 Belgrave Square
Dated: 21 November 1858
DDW 71/858/37

Pray answer me these queries.

Where have any old red fishes been found near Elgin exactly at Scat
Craig & the quarry of . . . . where Mr Robertson collected?[29]

---

[29]This was the Alexander Robertson who wrote the geological entries for Anderson's *Guide
to the Highlands* and after whom the *Stagonolepis robertsoni* has been named by Agassiz.

What are the species of *Holoptychius*[30] found there?

Are M^r P Duff's fish scales named specifically?

If not could you send us a few by rail. They will all be named & returned.

I am very anxious to settle this point.

The presence of the *Holoptychius* for the first time is teaching you the North is very important. Elsewhere this fish is unknown in the north, though abundant in the Central & Southern Counties of Scotland.

Mr Beckles is a thin skinned collector who fancies his rights of quarrying fossils have been interfered with. I have let him know that I took possession of the fossils there in your company before he went there & that the ground has been mine[31] for 31 years & now is yours.

I have no notion of the Elgin gentlemen & yourself in particular giving up their rights to an accidental interloper like Mr Beckles.

What is the distance in a straight line from the crystalline rocks S of Birnie to Lossie Mouth?

Where is the Findrassie Quarry—Is it in the ridge immediately S of Elgin?[32]

To: Gordon
From: [Jermyn Street]
Dated: 25 November 1858
DDW 71/858/38

I will entertain my friends at the next Geological with a copious dish of the Older Rocks of the Orkneys, Caithness Sutherland Ross[33] etc & do not intend to broach Morayshire til the next meeting ie the 15th Dcr, when I hope to do justice to it Geologically. Huxley will I doubt not astonish the [world] with the perspectives of the great Reptile *Stagonolepis Robertsoni*.[34]

Murchison had left a blank after the word "of." "Under the Knock of Alves" was the answer Gordon wrote on Murchison's letter. Whether Murchison received the answer is uncertain.

[30]In his letter to Gordon dated 16 November 1858, Huxley had already asked a similar question. Whereas Murchison continues to express an interest in the presence of *Holoptychius* in Moray in the hope of being able to demonstrate a continuity in the Old Red Sandstone, Huxley's purpose was to establish that the fossil remains of *Stagonolepis* had not been found in the Old Red Sandstone where fish scales had been discovered. The two letters obviously derive from a conversation of Huxley and Murchison's. Their enquiries would have developed differently had they gone together to the district that was now of such great interest to them, but they never did.

[31]Since his geological survey of the region with Sedgwick in 1827.

[32]See Map 1.

[33]The relevant Murchison papers at this stage were: i. "On the Succession of Rocks in the Northern Highlands, from the Oldest Gneiss, through Strata of Cambrian and Lower Silurian Age, to the Old Red Sandstone inclusive." *QJGS* 14(1858): 501–504. ii. "On the Sandstones of Morayshire (Elgin, etc) containing Reptilian Remains; and on their Relations to the Old Red Sandstone of that Country." *QJGS* 15(1858):419–436.

[34]Huxley had begun his paper on *Stagonolepis robertsoni* by acknowledging the assistance of Murchison, Gordon and Duff. See *Huxley at Work* pp. 36–43. What mostly astonished

The curious joint-like specimen which so struck me in M^r· Duff's collection, & which I carried away carefully in my portmanteau, has, I am happy to say, <u>recovered</u> the character which I first assigned to it of being <u>vertebral</u>. This I said to M^r· Duff. It now turns out to be no Cephalopod, as suggested at Leeds, but the <u>tail</u> of the *Stagonolepis*!!!

I believe I asked you in my last to send me any fishes (true or supposed fragments of) either from Scat Craig, Newton or elsewhere in the Elgin district.

The fact is, that my friends are sceptical as to the possibility of the *Stagonolepis* & *Telerpeton* being a true Old Red Reptile.

Independent of stratigraphical arrangement on which I am I think strong, I go much on the fact (on which I satisfied myself with you) that the Newton reddish sandstones with pebbles & many fish scales, do actually graduate up into the same yellow Sandstone as that which <u>as a whole ridge</u>, ranges from Spynie Castle & Findrassie on the ENE by the Hospital Quarry to Newton & the Knock of Alves on the WSW.[35]

If the fishes be really old red fishes & not scales & teeth of reptiles my cause is established.

Huxley's audience was not only that *Stagonolepis* was a reptile not a fish, but also that it was a reptile with crocodilian affinities. The idea that there had once been crocodiles around the Moray Firth shocked the Victorian imagination.

[35]This paragraph correlates with the field note-book M/N 134 of 1858 in the Geological Society Library, where Murchison records "a most satisfactory afternoon with the Rev. George Gordon of Birnie". On 6 September 1858, after examining the collection of Patrick Duff (much enriched since his last visit in 1840), he and Gordon "marched" from Elgin to the Linksfield quarry that has first been described by Lambert Brickenden in "On the Occurrence of the Boulder Clay in the Limestone Quarry, Linksfield, Elgin N.B." *QJGS* (25 June 1851): 289–92. Of Linksfield Murchison wrote: "I think there is here no case either for Capt. Brickenden or for my zealous friend Mr. George Gordon. The quarry appeared to have been simply thus":

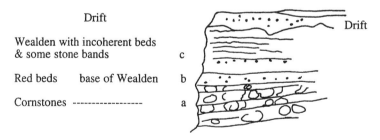

From Linksfield they went on to Spynie and Lossiemouth. It was on the next day, 7 September, that Gordon and Murchison set out westwards as far as Burghead, no doubt visiting Newton and the Knock of Alves en route, as mentioned in this letter. On folio 57 of this note-book Murchison wrote: "Notwithstanding the pains I took I now know that the Telerpeton sandstones are ... Triassic." This correction was written in May 1868. At the same time, Murchison wrote the word "omit" on each notebook page from 58–72, meaning "omit" from the transcription of the notebooks into the journal. Here, therefore, is an example of the journal *not* providing a record of what Murchison did during a particular visit. In retrospect, his afternoons with Gordon had not been as satisfactory as he had at first believed.

You will also <u>much</u> help by giving me a clear abstract of the points which Malcolmson made out—& lastly I beg you to tell me the name of the gentleman who has made the collection of Oolite & Lias fossils which <u>you</u> believe to be mere drift, or if there originally, no longer to be recognized <u>in situ</u>.

Where is the list of the <u>Wealden fossils</u> in the patch of stuff above the Cornstone near Elgin.

Did Mr Robertson describe these?

Lyell & all the world are perplexed by the *Stagonolepis*—in short we are all <u>staggered</u>. But I stick to my text.

I presume there is no doubt that the beds of Scat Craig are a continuation of the pebbly & <u>cornstony</u> red stuff which we visited & which came out on the road to . . . . .

What is the name of the place where the lowest red cornstone is—i.e. ENE from your Manse & the bridge whence we turned back to the high road.

Excuse all my trouble. But I wish to do justice to all & cannot effect it without your kind aid.

To: Gordon
From: Jermyn Street
Dated: 27 November 1858
DDW 71/858/40

Thank you for the transmission of the scales of *Holoptychii* which are unquestionably those of the upper strata of the Old Red. But alas they will not enable me to fight the fight for their <u>contemporaneity</u> with the fossils of the Findrassie & Spynie quarries in which Reptilian bones occur, without I can demonstrate that the beds of Bishops Mill, Newton, etc. are really part & parcel of <u>the same sandstone ridges of yellow & white colours</u>.

My opponents seeing the very high character of the Reptilian remains (the Findrassie & Lossie Mouth ridges contain the same things) may contend that these white sandstones are of Jurassic Oolitic? age & have been deposited on flat or slightly inclined beds of Old Red with *Holoptychii*.

Your rough map has a little section in it which if it could be maintained could quite establish my point. You place Bishopsmill higher than Findrassie thus.

& as the beds are only slightly inclined to the NNW, the inference might be that the beds with *Holoptychii* overlaid the Reptile beds.

But I apprehend that the reddish yellow beds of the Mill are in a <u>valley</u>, & lower than those of Findrassie, just as the Newton fish beds which I visited form the <u>base</u> of the ridge of the Knock of Alves.

Still, admitting this, I shall still hold to the white & yellow ridges of Elgin on the North & of Burg Head—Causie & Lossiemouth in the North (<u>for they are now clearly identified</u>) being upper portions of the Old Red, provided only the beds with *Holoptychius* <u>fairly pass up into them</u>. I also hold to this belief, because it seemed clear to me that cornstones marked the Lossie Mouth beds.

There are however people who may contend, that the whole series of cornstones & white & yellow sandstones belong to an <u>overlying [group]</u> of secondary age, reposing [   ] the undulations of Old Red with *Holoptychius*—that old [fraction] not extending northwards beyond the Elgin ridge & the younger deposits of like coloured Sandstone & Cornstone having been accumulated thereon.

In reference to these difficulties & finding from Huxley, that the *Stagonolepis Robertsoni* is nearly related to <u>Crocodiles</u> !! & knowing how obscure the country is, I must be cautious. The thing which chiefly staggers me is, that the fish scales are all those of *Holoptychius,* a genus unknown in the lower part of the middle & in the lowest beds of the Old Red Series & confined chiefly to the upper division—and that this fish occurs at Scat Craig, which I had supposed to be the oldest part of the Old Red of your tract.

The fishes of the Spey & Gamrie are [those] of Caithness & consequentially below the *Holoptychii.*

I have mislaid my old notes[36] & forget how the beds of Scat Craig dip— possibly they may be thrown over southwards & thus give us reason to hope for an <u>axial arrangement</u> by which the Banffshire & Findhorn strata might be represented in your great valley of denudation (in an arch).

Can you (as a fortnight occurs before my <u>hanging</u> day) revisit Newton & satisfy yourself if there be really what I [suppose] a passage upwards from the pebbly & fish beds into the yellow & white sandstones of the ridge—or if the same can be done—ie if you can any how unite the Bishop Mill strata with the same ridge & the Findrassie quarries.

In either case I shall stand firm to my case. If not I must be very cautious & [still] [leave] the whole case in doubt—

The worst of it is, that there is no where that I saw a single example where the Cornstones are [   ] with any other rock red yellow or white though my best recollection of 31 years ago (when however mineral char-

---

[36]On 9 September 1858 Murchison seems to have inadvertently left his notebook in his hotel when he went to visit Gordon at Birnie on the first occassion he had been to the manse. The notes he made that day when they visited Scaat Craig together have not yet been tracked down.

acters were dominant) led me to believe that the cornstones of the Find-horn are unquestionably <u>in</u> the fishy Old Red.

What says Malcolmson[37] on this point & does he also group the Elgin Sandstones with the Old Red.

In such a blind country (geologically speaking) I must have all the support possible so pray aid the blind man.

Yours sincerely
Rod I Murchison

It is so far a satisfactory confirmation of the accuracy of my geological coup d'etât that I insisted on the ridges of Elgin & the sea shore being synonymous before their included beasties were identified.

To: Murchison
From: Birnie
Dated: 30 November 1858
DDW 71/858/41

I am in receipt of your letters of 25 & 27, and do not regret to learn that your Morayshire paper is postponed until the 15 Dec[r], although I fear but little will appear before that time to settle the *questio vexata*—the age of the strata at Findrassie, Spynie, Lossiemouth & Cummingston.

From your note I infer that the scales sent last week prove that Scaat Craig, Carden Moor and Bishopmill are Old Red.

Bishopmill is the point that lies nearest to Findrassie, and my rude section was intended merely to show you the position of Findrassie as regards Bishopmill and <u>Elgin</u> (not the "Mill" as you seem to have read it.)

Although Bishopmill be the higher ground yet there is nothing to lead me to think that the Findrassie beds pass under it thus

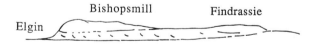

The reverse is rather the case,—namely that the Findrassie beds are the newer deposit, and rest on the Bishopmill strata. My supposition is that

---

[37]John Grant Malcolmson (1802–44), MD, FGS, FRAS, FRS. See Gordon's reply in the next letter.

the Findrassie bed is the same as that in which the *Telerpeton* was found at Spynie; and if the *Stagonolepis* bed at Findrassie was <u>underlying</u> the Bishopmill strata then these strata (at Bishopmill) would be the same as on the top of this hill at Spynie thus,

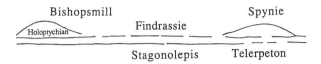

which does not appear to be the case, as we have not met with any *Holoptychius* scales at Spynie where also the sandstone is more compact & not so yellow as at Bishopmill. So I think we must admit that the Findrassie (a portion of the Spynie hill) beds have been deposited subsequently to those of Bishopmill. And, Bishopmill from its containing *Holoptychius* scales being unquestionably Old Red, the question is narrowed to—What is the age of all the cornstones & sandstones to the north of Findrassie from Burghead on the west, to Lossiemouth and Spynie on the east.—It is simply the occurrence of the <u>Reptiles</u> that has raised the question, as prior to this (their lithological character dip & strike agreeing with those in which *Holoptychius* scales are found) we were led to regard them as Old Red too; and even now there seems no other element than these Reptiles that makes one hesitate as to this same conclusion.

The vast accumulation of drift, covering the junction of the strata, is one great cause of this difficulty. Were it the case, as you say some may contend, "that the whole series of cornstones & white & yellow sandstones belong to an overlying group of secondary age", how comes it that so wide a range shews no fossils of this secondary age—Are any secondary formations so devoid of fossils as our cornstone—which being new yielded [   ] one yet not [   ] the absence of all carboniferous fossils. I do not much sympathise with you in the difficulty you suggest "that the fish scales are all those of *Holoptychius* a genus unknown in the lower part of the middle & in the lowest beds of the Old Red series, and confined chiefly to the upper division, and that this fish occurs at Scat Craig which I had supposed to be the oldest part of the Old Red of your tract."[38] I consider that the lowest fish beds (nodular) such as we have at Tynat[39] & Dipple on the east side of this district, & at Lethen bar & Clune on the west—if ever deposited have been afterwards swept away from the central part of Moray—particularly at the Scat Craig and the Findhorn where the fossiliferous beds lie near to the crystalline rocks. The fossiliferous beds (not the nodular fish) are largely developed on the Findhorn, and in breadth of outcropping surface would cover all the space from Scaat Craig to the Cornstone South of Elgin.

---

[38]We have not seen this letter.
[39]This location is spelled Tynet.

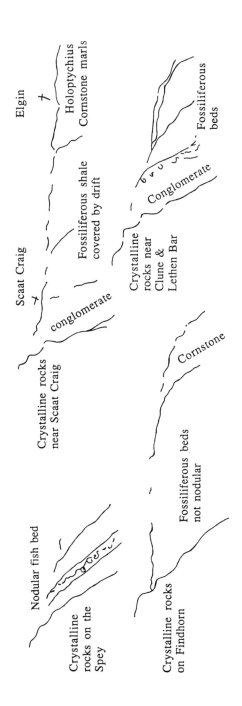

Elgin

Holoptychius
Cornstone marls

Fossiliferous
beds

Fossiliferous shale
covered by drift

Conglomerate

Scaat Craig

Crystalline
rocks near
Clune &
Lethen Bar

conglomerate

Crystalline rocks
near Scaat Craig

Cornstone

Nodular fish bed

Fossiliferous beds
not nodular

Crystalline
rocks on the
Spey

Crystalline rocks
on Findhorn

D^r. Malcolmsons paper suggests the last sections. I have never observed indications of any axial arrangement about the Scaat Craig or elsewhere so as [to] place the Banffshire and Findhorn strata at the opposite sides of a valley of denudation [   ]. Mr Duff quite agrees with you in believing that the red sandstones of Newton quarry graduate into the white or yellowish beds of the Knock of Alves & such is my own opinion but I propose to visit the locality soon.

Dr Malcolmson regarded the whole of our sandstones & cornstones as belonging to the Old Red—I have just corrected the last part of my paper on the history of the Geology of Moray for the Jany N° of the Edin^r Journal.[40] I am very severe on the Geolo. Society for not publishing Dr Malcolmson's paper. As what I have written contains a good deal of the D^r paper I have asked the Edin^r printer to send you the proof sheets.—I will be rather busy for a few days but shall endeavor to give you an abstract of Dr Malcolmson's paper before the 15th of Jan if the proof does not reach you.

The Rev^d Mr. Morrison of Urquhart[41] by Elgin is the gentleman who is forming a collection of Oolite fossils. I hope he will have a good set of them to exhibit at Aberdeen.[42]

I am not aware that Mr Robertson of Woodside[43] ever published any list of the fossils he collected beyond what he states in his note, which you read, in Anderson's Guide 3rd Ed^n.

Upon the whole, altho I would not at present declare for the Old Red out and out yet I hold a strong impression that it is so, and will not easily be brought to give up the hypothesis and would hold by this opinion until a *Stagonolepis* is found in some of the acknowledged newer formations of other localities, or until some Jurassic fossil be discovered at Findrassie, Spynie, Lossiemouth or in Mason's Haugh quarry at Cummingston.

I had a note from the tacksman (Mr Smith) of the Lossiemouth quarry that he had got some more bones. I went for them yesterday & if there be anything good in the blocks I shall forward them for you.

---

[40]"On the Geology of the Lower and Northern Part of the Province of Moray; its History, Present State of Inquiry, and Points for Future Examination." *Edinburgh New Philosophical Journal* (new series) 9, (January 1859): 14–58.

[41]The Rev. James Morrison, a noted local archaeologist and friend of George Gordon, became Minister at Urquhart in 1844 and Clerk of the Presbytery of Elgin in 1846. Noted for his "pawky" humour.

[42]At the Aberdeen meeting of the British Association in 1859.

[43]Of the seven Woodsides in the proximity of Elgin, this is the one in the Parish of St. Andrews, Llanbryde.

To: Gordon
From: [Jermyn Street]
Dated: 1 December 1858
DDW 71/858/42

Least you should think from my preceding letter that I am <u>shaky</u>
about the Old Red in consequence of the presence of a quasi Crocodilic
animal. I write two lines to tell you that on reconsidering all my evidences & particularly after referring to my section on the Findhorn in
1840[44] when I had first the pleasure of meeting you I am as firm as a rock
on the point.

The alternatives of red & whitish yellow marls near Altyre & up to
Cothall '<u>cornstones</u>' are so clear & decisive that all doubt vanishes.

Besides I am convinced it is impossible to put into different formations
the red rock of Newton & the . . . . . Mill[45] & the overlying yellow sandstones—or to divide the bottom red rocks at Lossiemouth from the overlying white beds.

I find I have detailed section of it & also of Scat Craig.

But the maps[46] are so infamous that if you could give me its relative position & particularly if you let me know if there be any other red Sandstone between it & the <u>crystalline rocks</u> of the interior you would much
oblige me.

The presence of *Holoptychius* there, shews that the bed is not very <u>old</u>—
& hence I want a place for the lower Member of the Old Red.

To: Murchison
From: Birnie
Dated: 5 December 1858
LGS/M/G10/1

I am delighted to learn that your notes of 1840 have turned up so opportunely
to confirm the views you lately took of the age of our upper rocks in Moray.

My last letter anticipated some of your queries. The only one remaining
to be replied to, is the position of Scaat-Craig. It lies 4 1/4 miles S.S.E. of Elgin.

From the sections I sketched you will see that the Scaat Craig beds lie on
a conglomerate (coarse), which again rests on the crystalline beds.

I hope Neill & Cº printers Edinburgh have sent you the proof of my paper, in which you will find Dr. Malcolmson's description of the fossiliferous beds on the Findhorn pretty fully given.

There is still time before the 15th, to forward any other information in
my power, if desired by you.

---

[44]We have not yet located this notebook.

[45]After the word "the" there is a blank. He may have again forgotten the name for Bishops
Mill.

[46]Notebook M/N 134 shows that he had consulted R.J.H. Cunningham's map of Scotland
and also his 1839 memoir in the transactions of the Highland and Agricultural Society of Scotland, Volume XIII. Notebook M/N 135 shows that he also used Patrick Duff's map in his
Sketch of the Geology of Moray.

To: Gordon
From: Ross Park, Petersfield
Dated: 5 December 1858
DDW 71/858/43

Your letter of the 28th Nov came just after my own of the 1st Decr. had left London & by which you would see that I had quite recovered my firmness of purpose in maintaining the Old Red age (Upper Old Red) of the Elgin deposits.

The only point on which I do not agree with you is that the nodular beds with the fishes of Caithness must, I think, be placed lower than the strata with *Holoptychius nobilissimus* of Scat Craig.

In order to satisfy myself on this point I presume that if they are persistent & were discernible those nodular beds would run along in that deep red concretionary band which we saw a cliff of in descending from the Schokel Falls[47] to the gorge in which the organic remains are found in yellowish soft sandstone.

By the bye tell me the name of that spot & also what were the fossils found there. Were they *Holoptychii*?

Owing to my absence from Town I have not got the proof of your Memoir on Malcolmson, but I know it is in London by a letter received that it is there & I beg to thank you.

In the mean time I have written out my own notice to precede the description of this terrible Crocodilean reptile which was to frighten us out of our footprints.

I care nothing for the high grade of the *Stagonolepis* except to admire the <u>beauty</u> of the oldest known reptiles whether *Telerpeton* & itself. As I am *toto coelo* opposed to the doctrine of transmutation of species so I rejoice in seeing the first created of its class as elaborately developed a reptile as the earliest trilobite of my Silurian rocks was as wonderful as any subsequent Crustacean.

As I know the Oolites of Sutherland better than any of my contemporaries I sweep away all the cobweb nonsense as to the possibility of their being of the same age as our coast ridge of Lossie Mouth & Burg Head.

I see by my old sections that Malcolmson found some of his fishes in flagstones near Altyre immediately overlying the bottom Conglomerate. Do you know what those fishes were. Perhaps this is explained in the Memoir you have sent me.

But as my day of Execution is the 15 I write to prevent the possibility of omissions which I might remedy.

I wish to know if Malcolmson or others obtained <u>fishes</u> in any of the red or yellow sandstones which on the Findhorn dip under the cornstones or limestone of Cothall—*Holoptychius* or others. According to my present

---

[47]Shoggle Falls. On a tributary of the Lossie, a short distance to the west of Gordon's manse.

view the yellow & red flagstones in which Lady G. Cumming[48] first and Malcolmson afterwards found the chief part of the fishes reposed immediately on the lowest conglomerate & sandstone.

Now those fishes were not any of them as far as I recollect of the *Holoptychius* zone, but of the same forms as occur at Lethen Bar—i e the Caithness Zone.

In the Findhorn section the beds with *Holoptychius* ought to occur somewhere about the Ramphlet Cliff[49]—i e in some of the pebbly calcareous beds like those of Scat Craig in your tract.

Pray tell me the name of the spot to the SE of Birnie in the hills where you showed me a red mottled rock which I said was a true Herefordshire Cornstone & which I believe to be under your conglomerate.

What is the maximum & minimum size of the fragments in your Birnie conglomerate which can only be a local expression of some of the similar beds on the Findhorn?—I have a note that the boulders or fragments in that conglomerate vary from the size of fists to that of man's heads.

Any crumbs you may contribute will be thankfully received by yours my dear Sir.

To: Gordon
From: Jermyn Street
Dated: 17 December 1858
DDW 71/858/46

Three days before the reading of my Memoir came off & that of Mr Huxley, I went to the Geological Society to have a regular search in the deposits & archives etc respecting poor Malcolmson's Memoir & every thing being in really good order (thanks to our excellent Assist Secretary) I found the Memoir all prepared for press (apparently) with notes in the author's hand written attached to the clean copy of his original Memoir.

Some of these notes are very valuable & refer to discussions of a fossil called erroneously I suppose *Cephalapsis* ?? *Gordoni* on the Nairn Shore.

Then came the question what stopped the publication & on referring to a regular record <u>book</u> kept by Mr Lonsdale it appears that the Memoir was referred to me—that I recommended it to be printed, but that it was not billetted for & ordered to be printed having been postponed until Mr Agassiz should have determined the specific character of the fishes!

After that I was entirely absorbed in my Russian Labour 1840 to 45— Hugh Miller & Agassiz making known all about those ichthyolites, & Malcolmsons Memoir (in which by the bye there are a few interpolations of D^r

---

[48]This mistake in priority was corrected by Gordon in his letter of 23 December 1858.

[49]See Map 2. At this locality the Findhorn cuts a deep gorge and the Old Red Sandstone is seen to lie nonconformably on Moinian rocks.

Falconer) passed into the Record Press where it was regularly packed up & forgotten.

The enclosed note I have sent to the Black's of Edin in the hope that they will use it as a fly notice in apology for the injustice done to Malcolmson.

It will be my duty now to see that all the original facts of his memoir be brought out & every credit done to his memory.

The Evening went off <u>admirably</u>. Everyone said it was the best Evening they had had for a long time.

I did my best both by positive & negative evidence (several big sections) & particularly by exhausting the arguments derived from Brora Devonshire etc to show that the Reptile Sandstones could not be part of that series.

<u>But</u> the Crocodilian relations of the *Stagonolepis* so ably made out by Huxley has I find made nearly all the geologists sceptical as to the geological order of Sedgwick & Murchison supported as they have been since by Malcolmson, Gordon, Duff, Robertson & Cº. & confirmed as far as he had the power of seeing and reasoning by <u>R.I.M.</u> (though no one dared to face me at the meeting)

Lyell spoke long & well & was of course <u>charmed</u> that so <u>high</u> a reptile occurred in so <u>old</u> a rock & congratulated me ironically & in a joking way upon feeling so comfortable at this introduction into that which I say cannot be anything but uppermost Devonian or lowermost Carboniferous seeing that it graduates downwards into the beds with *Holoptychii*.

I will write to you again when I see what can be done with Malcolmson's Memoir.

In the mean time pray send a line to Messrs Black if you approve of the notice which I have begged them to insert as a fly note in the coming number & with your Memoir.

Yours very sincerely
Rod I Murchison

If you prefer the printed document you may send it to Black & Co, who will be glad to take an excerpt for the forthcoming *Siluria*.

(Received 22 Decʳ. GG)[50]

The Memoir by the late Dr. J. Malcolmson on the Old Red Sandstone of Scotland has not been lost, but is still in the Archives of the Geological Society.[51]

Sir Roderick Murchison, having found it there and having referred to the Society's Books & seen the official entries respecting the Memoir in

---

[50]Murchison wrote this postscript at the head of the letter where Gordon made note of its receipt. Gordon must have received this letter on the 21 December from evidence in the letter of that date to John Hutton Balfour.

[51]Murchison may have forgotten that he had written to Buckland, then President of the Geological Society, on or near 26 May 1840, to urge that Malcolmson's paper be presented at the next meeting. With the letter he enclosed a formal report addressed to "The President" dated 22 May 1840 in which he said "The Memoir is highly worthy of a place in our Transactions." Murchison was not in London, however, to ensure that this was done. He was already on the Caledonian steamer on his way to Lubeck and from there to Russia. Malcolmson, too, had left London and was on his way back to India. Falconer reported on the paper as it had been

question, explained to the last Meeting of the Society on Dec. 15th, the causes of its not having been printed after he had, as Referee, recommended its publication. This delay arose it appears from the want, at the time, of descriptions of the fossil fishes referred to in the Memoir, as stated in the Secretary's handwriting the publication having been postponed until Prof. Agassiz should have described the specimens. By the time that Prof. Agassiz had described the fishes of the Old Red, the Memoir in question had been lost sight of, especially as Sir Roderick, for three or four years, was occupied with geological researches in Russia.[52]

To: Gordon
From: Travellers Club 6 ½ PM
Dated: no date[53]
DDW 71/858/47

A word to say that I <u>have</u> sent a copy of my printed document to Messrs Black & Co of the Edn Phil Mag saying that I much prefer having it printed to the MSS.

Yours sincerely
Rod I Murchison

I have added <u>in re</u> Malcolmson 'Every effort will be made to do honour [to] his memory'

To: Prof. Balfour,[54]
From: Gordon, Birnie
Dated: 21 December 1858
DDW 71/858/48/Balfour[55]

"In a letter just received, Sir Roderick Murchison informs me, that, after having read the proof of my paper (which was sent to him at my instance,

---

submitted, so it may have been he who recommended a deferral of publication until the new genera referred to in Murchison's letter had been named by Agassiz (Geological Society: Buckland/Murchison 18). Malcolmson never returned to Britain; he died in India in 1844. Murchison was busy with his Russian work for the next several years.

[52]These final two paragraphs, written on a separate leaf, are a copy of the note that Murchison asked the printer to include in his forthcoming edition of *Siluria*. With the autographed letter in the Elgin Museum is a proof sheet dated 17 December of the additional material for *Siluria*, London.

[53]Marked by Gordon "Rec^d. 20 Decr G.G." From the letter of the 23rd of December it is clear that this letter was written on December 17, 1858.

[54]Professor John Hutton Balfour. Editor of the *Edinbugh New Philosophical Journal*. Fellow of the Royal Society of Edinburgh, Regius Keeper of the Royal Botanic Gardens, and Professor of Medicine and Botany in the University of Edinburgh.

[55]Regrettably an up to date list of autograph manuscripts attributed to Malcolmson in the Falconer Museum does not exist. Some papers may not have survived the museum's care; an oil painting of Malcolmson is also missing. One of Malcolmson's Indian notebooks is, however, in the Elgin Library.

in order to supply him with some information he desired on our geology—prior to his paper on that subject before the Geol Society on 15th instant) 'a regular search' was made and Dr Malcolmson's memoir was found "all prepared for the press, apparently', and regularly packed up and forgotten," in the "Record Press" of the Geol Society of London.

Sir Roderick also tells me that he had sent a note of explanation to Messr Black & Co which he wishes published in the forthcoming N° of your Journal.

This request meets my entire consensus, and I trust you will kindly cause it to be complied with.

I am glad that I have been the means of bringing Dr Malcolmson's Memoir to light again. Trusting that in the circumstances you will forgive this additional trouble I now put you to.

To: Murchison
From: Birnie
Dated: 23 December 1858
DDW 71/858/48[56]

Immediately upon the receipt of your letter of 17th (which somehow happened to be 24 hours later than the arrival of the short one of the same date) I wrote to the Editor of the Edin New Phil Journal giving my entire concurrence to the insertion of your note of explanation in the forthcoming number. I am glad that I have been the means of bringing my excellent friend's memoir once more to light, and more pleased should I be to see it in print. I am given to understand that some of Dr Malcolmson's papers and specimens are still at Forres[57], which would likely further elucidate his labours. I must again press upon you my conviction that it was Malcolmson and not Lady Cumming Gordon[58] who discovered the fossils at Lethenbar. Mr. Stables of Cawdor[59], who was much with Dr. Malcolmson, assures me that this was the case. And Hugh Miller also, in the end of his VII Chapter of the Old Red Sandstone (p 168 3rd Edition) says, the seat of Sir Wm G. Cumming of Altyre is in the neighborhood of one of the Morayshire deposits discovered by Mr Malcolmson; and for the greater part of the last two years Lady Gordon Cumming has been engaged in making a collection of its peculiar fossils etc"—

A few more bones and plates will be forwarded to you, from Lossiemouth by the end of next week. In one of the bits there are materi-

---

[56]This letter has the same control number as the preceding one because it is on the same leaf.

[57]Possibly in Cluny Cottage, home of Malcolmson's mother.

[58]This inversion of Lady Cumming's names must have been a slip of the pen.

[59]W.A. Stables, who succeeded his father as factor or estate manager at Cawdor Castle. Gordon and Stables had been friends since their University of Edinburgh days and went on many expeditions together.

als (whether of tooth or bone I can't say) in forms such as that which Prof. Huxley said were "uncanny" looking things. Unless something very marked or new turns up, this will be the last parcel that I shall be able to send for some time, as the Elgin Literary Society now wish to have some specimens in the Museum to exhibit to the host of *Stagonolepis* hunters whom your paper will send down upon Murayland next season.

I shall however endeavour to secure a sight at least of anything really valuable for the purpose of Prof Huxley's completing the animal.—We are all delighted to hear that the evening of the 15th went off so admirably.

To: Gordon
From: 16 Belgrave Square
Dated: 30 December 1858
DDW 71/858/50

Before I start to spend a few days in the country with my friend Lord Broughton, I have the satisfaction to say, that I have corrected the last pages of my fat bantling 'Siluria' & have had time to put in poor Malcolmson before Lady Gordon Cummings as you wished.

My Dedication & all is done, & I shall send to you as my valued coadjuter in the Moray provinces a copy of the book (10 days hence).

I fear I cannot send copies to others concerned—for Siluria is not a book which alludes to your tract more than incidentally.

In the Memoir [of the] Geol. Soc^y. however on your County, special copies will be produced (printed separately) for M^r P Duff, yourself, Mr Martin & as many as you please to name.

I have taken care to extract the pith from M^r P Duff's interesting & characteristic sketch of the progress of research & discovery, & this will form a note in the said Memoir, which when bound up with Huxley's description of the Reptile[60], will, I hope prove satisfactory.

The <u>teeth</u> will be a capital addendum.

You must now let me have an account of your expenditure & I will then repay you with many thanks.

I am somewhat at a loss as to what ought to be done with Malcolmson's Memoir?? To print it *in extenso* after your Memoir will perhaps be more than the Council would like to do.

We might after an explanatory note as to the causes which stopped its issue at the proper season, publish the salient parts & take credit for him as to all <u>original points</u>.

---

[60]The paper by Murchison referred to in the previous paragraph is "On the Sandstones of Morayshire (Elgin, etc.) containing Reptilian Remains; and their Relations to the Old Red Sandstone of that Country", *QJGS*, 15, (1859): 419–436. The paper by Huxley is "On the *Stagonolepis Robertsoni* (Agassiz) of the Elgin Sandstones; and on the Recently Discovered Footprints in the Sandstones of Cummingstone", *QJGS*, 15, (1859): 440–460. See the discussion of the Murchison-Huxley confrontation in Part One, pp. 73–75.

You may calculate on a gay summer at Elgin if a geological foray will come into that category; for Lyell has told me he would like to be my companion.[61]

Remember me with many thanks to Mr. P. Duff & assure him that I have written a strong paragraph expressive of my regret, that after being the real possessor of the <u>two</u> Elgin reptiles (which without his liberal & perspicaceous encouragement might never have been preserved) neither of them have been named after him!

Reply to me so that I have your letter on or before Wednesday night, when our next Council meets, & when I wish to have some measure taken about Malcolmson's Memoir.

With the heartiest wishes of this season to you & yours & my remembrances to M^rs & Miss Gordon believe me to be

My dear Sir
Yours very sincerely
Rod I Murchison

Would it not be well to send Mr Martin £2 or £3 for his bone?

I do not like to take it for nothing from a poor man when richer people sell their fossils—

To: Murchison
From: Birnie
Dated: 1 January 1849[62]
DDW 71/859/1

I have just received your valued note of Thursday, and hasten to reply— as you wish for an answer before Wednesday—as regards Dr Malcolmson's paper.

I think what you suggest would, in the circumstances of the case be very proper, but before the principal or any part of it is printed, I should wish to have an inspection of his manuscripts—that I believe lie at Forres, which might give matter for a short memoir of the author & of his geological labour elsewhere. This would bring him before the Geological Society—in somewhat fresher attire than if the notice were confined entirely to his old paper. I shall apply for access to the repositories[63] at Forres as soon as I get one of the separate copies of my paper to send to his brother in London. If my request be granted I will report to you what I find.

You ask me to state what I have been out of pocket for the remains of the *Stagonolepis* forwarded to you. £5 will cover all.—More than half of

---

[61]We have no evidence to confirm that Lyell and Murchison visited Elgin at the same time, although both were there in 1859.

[62]Although this letter was dated 1849, it was certainly written in 1859 as its contents make clear.

[63]Since the Falconer Museum was only built in 1871, we do not know what repositories are being referred to.

this, by making no bargain with him, went to the Burghead carpenter for the boxes for the footprints, then packing & carrying them to the steamboat office. Contrary to Mᵣ Beckles experience wood was to be found at Burghead, altho not in abundance if we are to judge from the value put on it.

Be assured the Directors of the Elgin Museum are as anxious as I am that Profᵣ Huxley should see and examine anything new that shall now turn up in the quarries. All look to him as the man who is fully to reconstruct the *Stagonolepis* and every material for doing so they wish lay before him.

Any of the sums you mention would be very acceptable to Martin and would enlist his future exertions in the service.—

I could dispose warrantably of 8 or 10 copies of your Memoir upon our local geology—to individuals who would highly appreciate & profitably study it.—

Mᵣˢ Gordon & my daughter are much gratified by your kind remembrance of them and unite with me in warmly reciprocating your compliments of the season.

To: Gordon
From: John Anderson[64]
Dated: 12 January 1859
DDW 71/859/3

I have many thanks to give you for the much valued specimens of *Stagonolepis* you have so kindly sent me as well as for your former attentions through my esteemed and active friend Mr. Mackie[65] of Elgin.

The *Stagonolepis* I greatly prize[66] and the more so that they are now pronounced to be Crocodilia. Will you not have to shift the geognostic position and relation of the Sandstone Matrix? <u>Mineralogically</u> it looks not to my observation like our yellow sandstone whatever place it may turn out to hold in the system.

---

[64]This was the Rev. Dr. John Anderson who presented a paper to the Aberdeen meeting of the British Association entitled "On the Dura Den Sandstones." His publications included *The Course of Creation* (1851); *A Monograph of the Yellow Sandstone and Remains in Dura Den* (Edinburgh, 1859); "The Geology of Age," "Man in his Present Aspects" and "The Geology of Scotland" in *Pictorial History of Scotland* (1852) and "Geology of Fifeshire," a gold medal prize essay published in the *Transactions* of the Highland Society in 1838.

[65]This must have been the father of the future Medical Officer for Health in Elgin.

[66]The dispersal of reptilian fossil specimens is referred to in Part One and in *Huxley at Work*, where it is argued that the receipt of fossils in London depended partly on Huxley's willingness to return them and partly on whether or not Gordon felt that Huxley was displaying sufficient interest in what he sent him. Huxley did not write to Gordon between 30 December 1858 and 19 May 1859. During this period, and also at other times, Gordon sent specimens to those who did express interest, as here. Probably Huxley did not know that Gordon was doing this. Research was thus jeopardized when the chance to bring all the palaeontological evidence into one place was lost. This in turn delayed the moment when Murchison would *have* to take the palaeontology seriously.

I shall not fail to forward you in good time a right good specimen of Dura Den. The slabs are generally large and easily broken—which renders it the more difficult to make a proper selection for package and transmission.

I suppose you will be all in the north making great preparations for the Aberdeen Meeting and fights. Sir Rod[r]. gives up nothing of his new gneiss theory[67] in his forthcoming Siluria—

To: Gordon
From: Jermyn Street
Dated: 10 February 1859
DDW 71/859/7

I have been remiss in not sending you before now a copy of my 3rd Ed[n] of Siluria.

Accept one which goes by this days post & put it on your shelves at Birnie, but not before you have read great parts of it besides my North Scottish episodes & appendices.

'Rome was not built in a day', & certainly the true age of all the native strata which have been so metamorphosed will not be easily worked out.

As one of the hardy pioneers in this reformation of our old rocks I lay before you what I have endeavoured to do.

I have not yet received from you the accounts of your <u>expenditure</u> in procuring the flagstones.

After all Huxley attaches less value to them than I thought he would— a <u>bone</u> or a <u>good cast</u>, is gold for him; but the footmarks are not sufficiently decisive of anything more than that they were the impressions left by old & young reptiles—probably the *Stagonolepis*.

Rod I Murchison

PS As they wish me to give a sort of Evening Address at the Aberdeen Meeting on the Geology of the Highlands, I must of course get up & render my case <u>much more</u> perfect than it is.[68]

I intend therefore (DV)[69] to revisit Morayshire before the Scientific Tryst as well as many other places.

To: Murchison
From: Birnie
Dated: 14 February 1859
LGS/M/G10/2[70]

I beg very gratefully to acknowledge your valuable and much esteemed present-a copy of 'Siluria' duly received by me on Friday. I am as proud

---

[67]For a detailed discussion of Murchison's ideas about the origin of the Moinian rocks of the northwest Highlands see Oldroyd 1990.

[68]The text of this address is reproduced in Appendix 5.

[69]*Deo volente.* God willing.

[70]The draft of this letter is DDW 71/859/8. The fact that the letter as sent and the draft in the Elgin Museum are almost identical to a large extent establishes the status of the other

of it as of any gift I ever received, & shall study it with profit, & cherish it with care.

The Directors of the Elgin Museum have collected many specimens from the <u>debris</u> at Findrassie. I am not able to detect among them either a skull or tooth cast, else I should have endeavoured to have got it forwarded to Prof. Huxley. The <u>bed</u> in the quarry, whence these remains must have come, has not yet been seen by us. I am under the impression that it lies in the bottom of the quarry under rubbish, broken from the <u>upper siliceous strata</u>—similar to the relation of the Lossiemouth beds.

The Directors are also to attempt an excavation at Lossiemouth—where I hope something that will assist towards the completion of the huge reptile will turn up. And if the latter locality yield the same amount as Findrassie, I do think the better plan would be for Prof[r]. Huxley to see[71] them *in cumulo,* than for any inexperienced hand to select a portion of them for transmission to London. We might overlook what was valuable and send merely repetitions of what he has already seen. To send the whole mass would be out of the question—so it will likely come to the old resolution of the dilemma,— 'Mohamet to the Mountain'. However I shall see what can be done.

Observing at Cummingston footprints of animals apparently as small as the *Telerpeton* among those of a larger size, a question has suggested itself to me, which I have no doubt Prof[r] Huxley could at once answer—viz. Are there distinctive marks ascertained sufficient to show that the *Telerpeton* is not the young of the *Stagonolepis?*

M[r]. Young told me the other day that M[r]. Anderson (of Cummingston) had discovered footprints in the Clashach quarry at Covesea, which you recollect looking down upon from the top of the cliff.[72]

D[r] Taylor who had sent some labourers to search for fossils at Pluscarden showed me today an impression of a *Holoptychius* scale which they had found there, —the only vestige of an organism I could see among a lot of stones from that locality. Lithologically the isolated sandstone of Pluscarden resembles that of Quarrywood.

D[r]. Malcolmson's brother has written me from London giving full and hearty permission to examine the geological papers & specimens at Forres—which I mean to do soon.

You ask me as to my expenditure. Looking at my memoranda, I find that Five pounds (£5) will cover all my outlay for boxes and quarrymens wages etc. I wish I could have <u>franked</u> the whole.

---

drafts reproduced in this book. Both the letter and the draft are therefore reproduced here in order to establish the fact that in the majority of cases the two are close enough to being identical for the draft to be trusted.

[71]Notwithstanding this advice, Huxley remained unwilling to do his own field work, partly because he lacked the time, and partly because of his adherence to the methodology of Cuvier where a palaeontologist had only to have parts of a skeleton in hand for the whole to be reconstructed.

[72]It is difficult to know whether or not this was a mildly barbed remark. Murchison never devoted quite enough time to Elgin sites, something which Gordon had obviously noticed.

I am delighted to think that Morayland will again engage your attention during part of the coming autumn. I shall be on the outlook for some fresh points of interest among our drift covered rocks.

Again warmly thanking you for the 'Siluria'.

To: Murchison
From: Birnie
Dated: 14 February 1859
DDW 71/859/8

I beg very gratefully to acknowledge your valuable and much esteemed present—a copy of 'Siluria' duly received on Friday. I am as proud of it as of any gift I ever received, shall study it with profit, and cherish it with care.

The Directors of the Elgin Museum have collected many specimens from the debris of the Findrassie quarry. I am not able to detect among them either skull or tooth, else I should have endeavoured to get such desiderata forwarded to P Huxley. The bed in the quarry whence these remains must have come has not yet been seen by us. I am under the impression that it lies in the bottom of the quarry under rubbish—broken from the upper siliceous strata similar to the relation of the Lossiemouth beds.

The Directors are also about to excavate at Lossiemouth where I hope something that will assist towards the completion of the huge reptile will turn up. If that locality yield the same amount as Findrassie I do think the better plan would be for Prof^r Huxley to see them *in cumulo*, than for any inexperienced person to select a portion of them for transmission to London. We might overlook what was valuable & send merely a repeti-tion of what he has already seen. To send the whole would be out of the question—so it will likely come to the old resolution of this dilemma—Mahomet to the mountain—However we shall see what can be done.

Mr Young told me the other day that Mr Anderson of Cummingston had discovered footprints in the Clashach quarry at Covesea which you recollect looking down upon from the top of the cliff.

D^r. Taylor, who had sent some labourers to search for fossils at Pluscarden, shewed me today an impression of a *Holoptychius* scale which they had found there—the only vestige of an organism I could see among a lot of stones from that locality. Lithologically the isolated sandstones of Pluscarden resembles that of Quarrywood.

Dr Malcolmson's brother has written me from London giving full & hearty permission to examine the geological papers & specimens at Forres,—which I mean to do soon.

You ask me as to my expenditures. Looking to my memoranda I find that five pounds will cover all my outlay for boxes & remuneration to the quarrymen & other labourers I employed.—I wish I could have franked the whole.

I am delighted to think that Morayland will again engage your attention during part of the coming autumn. I shall be on the out look for some fresh points of interest among our drift covered rocks.

In observing at Cummingston footprints of animals as small as the *Telerpeton* among those of a larger size—a question has suggested itself to me which I have no doubt but P Huxley could answer at once— viz. Are these distinctive marks sufficient to show that the *Telerpeton* was not the young of the *Stagonolepis*.

Again warmly thanking you for the Siluria.

To: Murchison
From: Birnie
Dated: 7 March 1859
DDW 71/859/11

My last note was in acknowledgement of your handsome present of 'Siluria'; and I now fulfill a promise then made to report any occurrence calculated to throw light on the geology of Moray.

On this day week I heard that Mr. Smith had come upon some remains in the Lossiemouth quarry where you recollect getting the bones of the *Stagonolepis*. I went down and brought all I could collect to the Museum at Elgin, where I succeeded in putting a few of the fragments together. Today I revisited the locality, but nothing has appeared, altho the workmen have been quarrying in a very likely place. Without further delay then I make an attempt to represent very roughly by figure of <u>the most striking parts</u> of these organisms.[73] Four of the pieces when put together shew, as in the accompanying rude sketch what appears to me to be the skeleton of a <u>flat osseous fish,</u> which I fear for the Old Red would be proving too much! The two small bits of stone enclosed contain some of the streaks of the bony matter which I take to be ribs from the vertebral column.—Larger masses of ill defined bony matter lie in two or three other blocks which I cannot dovetail into, but which, from what the quarry man said, I believe to belong to the same animal. Throughout the whole mass nothing like a scute or the impression of one could be seen—or indeed of any osseous scale such as the Old Red Sandstone fish have so frequently.

PS The spine like processes coming off from what I call the vertebral Column may appear more numerous than they really are owing to the way in which it is broken up.

---

[73]Deleted from the text at this point "which were found all together."

To: Gordon
From: Jermyn Street
Dated: 11 March 1859
DDW 71/859/12

Thanks for your last communication which has interested me much. But before I talk about it I beg to enclose two orders for £5 each—the one in payment of your slabs of stone with reptilian footmarks And—the other for good Mr. Martin's bone or rather bones; for without his we would not have hit on the others.

You & he have only to date & sign them over a penny stamp & the Elgin banker will cash the order.

I am not alarmed by the appearance of your 'flat fish'! On the contrary if a guess of mine be worth anything I would opine, that the long-<u>undivided</u> vertebral column, (without joints) has a mighty Paleozoic aspect.

Owen gave the first of a course of lectures on fishes yesterday[74] the last 10 of which will be on fossil fishes.

As one of the existing singular types of fishes [that] had reference to Palaeozoic Fishes, there was hung up the drawing of the skeleton of a *Lepidosiren* as inhabiting African rivers, & I immediately fancied I saw an analogy with your fossil.

Huxley is somewhat unwell & has not been here for 2 days; so I could not consult him as to my supposed analogy & I did not like to talk to Owen about it; as I have placed all the Elgin fossils under the former (there being great rivalry between them) & as he has given us such a good account of the *Stagonolepis*, I think it right that he should have <u>the first peep</u>. I did however show your drawing to Lyell who was much struck at my analogy.

So that I should be very proud if I am right & that more as it will keep 'flat fish' if not <u>in</u> Old Red, at all costs <u>out</u> of the Secondary rocks.

Say, however, nothing about this until I know more & pay for me (if you think it right) a sovereign to the man who found these remains & promise more if he or his companion find other things. Once upon the scent & on the same horizon as that of the Stagonolepis, <u>pray keep up the cry</u>.

The determination of the question is intensely interesting.

Like your fossil the *Lepidosiren* has <u>bones</u> (large ones too) near the head only & has no hard scales.

After the signature:
I have sent Mr. Martins cheque to himself.

---

[74]The course of lectures referred to here was the series offered on fossil fish at the Museum of Practical Geology in Jermyn Street during the spring of 1859 (see Gruber and Thackray 1992 with its references to the printed synopsis in the Owen Collection of the Natural History Museum, OC38.3/282; and Rupke, 1994, Table 3).

To: Murchison
From: Birnie
Dated: 15 March 1859
DDW 71/859/13

I beg to acknowledge receipt of an order for £5 as payment in full of my expenditure to carpenters, quarrymen, etc.

Yesterday I saw the directors of the Elgin Museum in whose keeping the specimen which I brought from Lossiemouth now is, when at my request they readily consented to allow me to forward it for your inspection, provided it be returned to Elgin before the meeting of the Brit. Ass[n] at Aberdeen, along with the other specimen belonging to the Elgin Museum now with you in London. Prof[r] Huxley indeed deserves the "first peep" as you say.

I promised that the return of all specimens belonging to the Elgin Museum would be duly attended to; and I am sure you will the more readily help me to make good my word to my brother-directors as we wish to comply with the pressing solicitations from Aberdeen for specimens to fill temporarily the as yet empty cases of the Exhibition room during sitting of the Association.

It might be well then to arrange so as to have them all brought to Aberdeen where you could select such of them as will be suitable for illustrating your evening address on the geology of the North of Scotland.

I shall therefore forward without delay the blocks of stone containing what looks so like a flat osseous fish from the *Stagonolepis* bed of the Lossiemouth quarry and will be on the tiptoe of expectation to learn what it is thought & said to be by Prof. Huxley.

It would be superfluous at present to give the quarrymen more money—as they are sufficiently aware of the profit arising by their getting hold of these organisms to which their eyes are knowingly open—specially as the tacksman[75] is promised assistance from the Museum to clear away rubbish lying on a bed in which we know there are organic remains. I was at the quarry yesterday but nothing has been seen lately. G.G.

PS. 5 P.M.

I have packed up & given to the rail a box containing the organism from Lossiemouth. It will likely be in Jermyn St. about Saturday.

---

[75]Tacksman. One who holds a tack or lease of land. Esp. in the Highlands, a middleman who leases directly from the proprietor of an estate. *OED*

To: Murchison
From: Birnie
Dated: 12 April 1859
DDW 71/859/18

Some weeks ago I forwarded to you[76] a box containing the fossil found at
Lossiemouth of which I had previously sent a rude sketch. I know your
every hour is fully occupied but two lines to say whether or not it reached
Jermyn St will satisfy me & the other Directors of the Elgin Museum that
their specimen is in safe keeping or will set us on the search after it among
the Railway and Steam Cos through whose hands it was to pass.

    Beyond finding another locality[77] about two miles due west of this from
our lowest fossiliferous beds (same as Scaat Craig & Sheffleburn[78]) nothing new has turned up in the Old Red [sstn] for the last few months.[79]

To: Gordon
From: Jermyn Street
Dated: 14 April 1859
DDW 71/859/19

No amount of occupation could have prevented my replying to your letter announcing the transmission of the newly discovered fossils.

    The delay has been solely caused by the doubts of Prof Huxley con-
cerning the fossil bones you sent up from the Elgin Institution.

    He discards them from Fish & adopts them as Reptilian.

    But as he is not here today (Mrs H having just been brought to bed of a
girl) I must say no more & not compromise him too far.

    All these & the other remains shall be transported to Aberdeen where
they will be exhibited as the produce of the industry and research of the
fossilists of Morayshire. Before that time I intend D.V. to have a good skirmish over your grounds & to devote a week at least to a survey of all the
doubtful points.

    Huxley will write on the subject of the fossils.[80]

---

[76]The phrase "at Jermyn Street" deleted at this point.

[77]Here the phrase "at Pluscarden" was deleted. Gordon may have felt that the exploration
around Pluscarden in some sense belonged to other people and therefore could not be shared
with the London experts.

[78]Sherrifburn.

[79]He also deleted a final sentence about going to Forres to look at Malcolmson's Scottish
specimens.

[80]An addition to the original letter was made to explain the missing line and signature, as
follows: "His signature cut off (1869) and given to Miss [   ] Burgie GG"

To: Murchison
From: Birnie
Dated: 18 April 1859
71/859/21

At the end of my hurried note of Friday last, I mentioned that I had just heard from Mr Smith, the tacksman of the quarry[81], that he & his workmen had come upon some more fossils. I went down early on Saturday and obtained from them what, when put together, forms an admirable specimen of the last reptile I sent. Although I got one, of the large blocks that contain the animal, in the <u>quarry</u>, and the other at a building (a house or dyke) to which it had been carried, I am persuaded that they had originally been in juxtaposition. The vertebral column, with a multiplicity of <u>ribs</u>?, on one of the sides, and some of the bones of both extremities are very distinct. The head jaws and <u>teeth</u> (I am sure of them and cannot be mistaken altho small in proportion to the size of this animal).

After inquiry of the quarrymen, I saw that this specimen is from 8 to 10 feet higher up than that which has hitherto yielded fossils. This gives us hope that others will the sooner be found—their range, in depth of the strata, being thus enlarged. The specimen belongs to the Elgin Museum, the Directors having very helpfully engaged the services of Mr Smith and his workmen. All who take any great interest in such matters, having seen and speculated about this rarity, I readily got the sanction of the Directors to forward it for your and the inspection etc. of Prof. Huxley[83].—I have therefore packed it up and given it to the Railway Co. It should be with you about Saturday or Monday first, when I am certain it will gratify you much.

I shall feel obliged by a short note announcing its safe arrival in Jermyn Street, in the first place, and in the next by a report of Prof Huxley's views regarding it which we are most anxious to learn.—It would be de-

---

[81]At Lossiemouth.

[83]Huxley wrote to Gordon on 23 April 1859 (DDW 71/859/23), and what now follows is the whole of his important letter.

"I write a hasty note (to save the post) to tell you that your last shipment arrived quite safely.

I have put the fragments together & the hasty inspection which I have made of them is sufficient to enable me to say that the creature <u>is not</u> *Stagonolepis* & <u>is</u> one of the most remarkable anomalous Reptiles I have seen. Pray continue your search. The present specimen resolves many doubts about the first & every fragment is of value.

I shall be able to say nothing decided about the creature for some weeks. There is nothing for it but to sit down knife & chisel in hand, before the fossil for a week or two & fairly compel it to reveal its secrets by dint of scraping."

This marks the beginning of Huxley's work on *Hyperodapedon*, work which was to have such a serious impact on Murchison. But for the time being Murchison did not publically interest himself directly in Huxley's scraping, while Huxley's paper on the *Hyperodapedon*, after which the conflict between the geologists and palaeontologists could not be ignored, did not appear until 1869, that is to say, after the publication of the fourth edition of *Siluria*.

sirable that the distinction, if any, between it and the *Telerpeton* should be established for its ribs? are rather *Leptopleuronic* (Owen's name) "thin ribbed"—

I have marked the small bits so that they can be readily put in their proper places. The larger block, particularly that which contains the head will require to be broken up to get fully at their contents.

It gives me pleasure to think that you will be able to devote some days this autumn to the elucidation of our now so highly interesting geological locality.

To: Murchison
From: Birnie
Dated: no date[84]
DDW 71/859/58

Some ten days ago, accompanied by my friend Mr. Stables of Cawdor, I went to Lossiemouth and by the assistance of the quarrymen we obtained large blocks with bone scutes etc. of *Stagonolepis* from that part of the quarry whence the Directors of the Elgin Museum had engaged & paid M[r] Smith the tacksman to remove a large quantity of debris. Next day D[r]. Taylor H.E.I.C. superintended the removal to the Elgin Museum of three masses which we were unable to transport to the Museum.

Under the auspices of the Directors, D[r] Taylor & I yesterday filled a large box with some of the best pieces together with some from Findrassie and addressed it to you. When you shall have seen them I suspect you will say Enough. It will likely reach Jermyn St. by Saturday, when I trust you or PH. will kindly inform us of its arrival—Among them are the <u>teeth</u> at last and I trust other portions which will enable P[f] Huxley to complete his description of the animal.

Naturally desirous of making the Aberdeen meeting as much as possible the occasion of creating & spreading an interest in all that regards the Geology & Natural History of the North, I am requested by the Directors to say that they fondly hope that Prof. Huxley will kindly take the opportunity to give to the Geological Section some further account of these remarkable & unlooked for reptiles and they confidently trust to you that these and their other Specimens will be exhibited in the Granite City am-

---

[84]Huxley's letter of 19 May 1859 must be the answer to this, since Huxley says that he is going to the Aberdeen meeting of the British Association "for the precise purpose of giving a 'screed' about your famous Reptiles." What Gordon here calls "a large box" Huxley in the reply calls it "your last package" and then says: "It is a *Lacertian* having some relations with *Placodus* with *Rhynchosaurus* with (alas for Sir Roderick!) unquestionably Triassic forms." Huxley has dubbed this new reptile *Hyperodapedon Gordoni.* Meanwhile, though still in London during May and the early part of June, Murchison was preparing for his summer expedition. This undated letter must have been written in mid-May 1859, several months before the Aberdeen meeting of the British Association.

ple proof as you are pleased to call them of the industry and research of the fossilists of Morayshire.

I have written a few lines to Prof—Huxley regarding the diff—parcels in the box.

On visiting the sea side quarries last week I found that the new locality for the footprints was Greenhaugh & not the Clashach quarry as I previously referred you. The Greenhaugh is at least a mile distant from the old locality—

To: Gordon
From: Belgrave Square
Dated: 12 June 1859
DDW 71/859/30

I send you herewith the slips of my Memoir—or Part 3 of my North of Scotland concerns which are to be published on the 1st Aug with a General Map of the N of Scotland & Sections.[85]

I beg you to read this part & return it to me with any corrections or suggestions you may please to make & <u>quickly</u>.

Your last discovery of the *Hyperodapedon* of Huxley has so affected me, that I [am] I am a good deal shaken in my opinions as derived from the stratigraphical appearances & I propose to make a much longer foray in Morayshire than I did last year.[86]

I will also reexamine all the yellow Sandstones of Tain & Tarbet Ness in Rosshire, which, as well as those of the Findhorn, seem to me to be unquestionably interbedded with the Old Red. But, (as I have put it in the note appended & trust you will return with the Slips) it is just possible that we may be able to find out some signs of <u>transgression</u> between the fish beds of the Old Red & the Sandstones of the Elgin Ridge??

I regret that I did not give a <u>whole day</u> to the Newton case—i e in tracing the passage from the fish beds upwards. It certainly seemed to me that all the ridge there was one [physical] mass.

Possibly we may never, in your blind country, get any better evidence than we possess, but we must work hard.

I shall also revisit the Findhorn tract & work up to the white sandstones of Nairn. As all the best parts of Dr. Malcolmson's paper will be printed & will precede my Memoir, the note alluded to will be there inserted—In fact an Introductory [   ] showing how it was that the omission occurred.[87]

---

[85]*QJGS*, 15, (1859): 419–421, Plate XII. This is the precursor to the map of Scotland published with Geikie in 1861.

[86]On the expedition of 1858 Murchison was accompanied by Charles Peach. In 1859 Murchison traveled with Ramsay. He in fact was to spend two weeks in and around Elgin.

[87]Murchison's explanation of the delay, initialed by the editor, was published in *QJGS*, 15, (1859): 336.

To: Murchison
From: Birnie
Dated: 16 June 1859
DDW 71/859/31

I return the proof and M.S.[88], with a perusal & correction of which you have honored me.

Of all my <u>pen and ink</u> corrections I am pretty sure save that of "stria" into "strata" near the bottom of page 5. Those <u>in pencil</u> are given as suggested additions & improvements. I have also enclosed a few notes in a separate paper.

I have not the slightest fear of ever being induced to give over the body of the Quarrywood ridge = (that which runs from Alves eastward to Bishopmill quarry, inclusive however yellow or white its beds) to any <u>other</u> formation so long as *Holoptychius* is to be held characteristic of the upper division of the <u>Devonian or Old Red</u>; for this scale is found pervading the whole southern slope in which quarries have been opened.[89] The field of battle will be between the <u>Bishopmill</u> beds & those of <u>Findrassie</u> and Spynie.

It will delight me to give further explanations in case my emendations have had the effect of increasing the difficulty. I trust the whole will be made plain on your promised visit, before which I shall be able to conduct you to any spot where the rock crops out between the Findhorn and the Spey.

To: Murchison
From: Lyell
Dated: 21 July 1859
M/L17/29

I have been looking over the numerous & effective illustrations & reading the Appendices & some other of the new parts of your 3rd Edition with much pleasure & interest & liked much the dedication & Preface—It was a good bit to get Owen's note on the Conodonts the molluscous nature of which when I first saw them I strongly suspected.

What you have said of the *Holoptychius* not going up into the reptiliferous sandstone of Elgin leaves the door open to a future possible correction of the age of the *Telerpeton* & *Stagonolepis* tho' at present the evidence seem satisfactory.

But we want more proof than usual in such a case—I shall go over the whole with care when I have got my volcanic paper off my hands meanwhile many thanks for your kind present & written dedication at the beginning.

---

[88]Malcolmson's article with Murchison's introduction.

[89]Gordon deleted a new quarry location in a phrase as follows: 'and even in a new quarry near the middle of the ridge and'. It seems reasonable to suppose that Gordon was cautious when it came to identifying new sites because he knew of other people's interest in them.

To: Gordon
From: Brahan Castle, Dingwall[90]
Dated: 25 August 1859
DDW 71/859/28

Having reexamined every rock in Sutherland & certain points in Ross (the latter as tending to red, yellow and <u>white</u> Sandstones of Tarbat Ness) I am now wending my way to Moray.[91]

On Saturday I go to Lord Cawdor if they can receive me & Prof. Ramsay[92] (who is with me).

On Monday & Tuesday I shall be examining the Findhorn & adjacent Sections & hope to be at work with you on Wednesday from Elgin.

I trust that in the course of the week I shall have satisfied myself on the critical point of the Reptiliferous Sandstones. But I must let Ramsay (who I bring as a sort of Umpire) see the Lossie Section from the high country.

I must give up a whole day or more to the Coast section, & must spend more time upon the relation of the rocks at Newtown and thence Westwards to the Knock of Alves.

If you receive this tomorrow send me a line to P.O. Inverness so that I may get before I go to Cawdor Castle.[93] I am happy to tell you that Ramsay quite confirms all my views respecting the Old Succession.[94] I have nothing to change or even modify. The Reptiles alone bother me—but tell Mr P Duff he begs me to adhere to the old doctrine.

---

[90]Brahan Castle was the ancestral seat of the chiefs of the clan MacKenzie which, however, had suffered a series of setbacks since Donald Murchison, brother of John Murchison, Roderick's great grandfather, had been factor to the Earl of Seaforth. See Norman Macrae's *The Romance of the Royal Burgh* (of Dingwall). The castle was razed to the ground some years ago.

[91]The letter of 12 June 1859 made it clear that visiting Moray was, if not the, at least one of the principal reasons for his summer expedition that year. Oldroyd makes it seem as though, once Murchison and Ramsay had completed their work in the North-West, they hurried on to Aberdeen. "We need not concern ourselves", he said, "with the observations of the latter part of the journey" (p.86), thus not treating the work Murchison did between Eriboll in the North West and Aberdeen. The actual itinerary and the time it took can be reconstructed from the relevant field-notebook (LGS/M/N135) as has been demonstrated.

[92]Andrew Crombie Ramsay (1814–91). Scottish geologist. Appointed to the Geological Survey in 1841; professor of geology in University College and the Royal School of Mines; succeeded Murchison as Director-General in 1871. FRS 1849. Knighted 1881.

[93]Murchison received the Earl of Cawdor's invitation dated 25 August as he passed through Inverness, where he was to take a train to Cawdor Castle; "King Duncan's room shall be prepared for you and I hope you will pass a better night than he did in it". (M/C40/6).

[94]It is worth noting that by the time this letter was written, Murchison's relationship with Archibald Geikie, of importance later, was strengthening. In a letter to Geikie which refers to his 1859 expedition, he said: "The last three years of my life (as regards my own career as a geologist) have been chiefly spent in preparing and working out a new classifcation of the rocks of my native country, the Highlands of Scotland, and I am now endeavouring to give the last touches to this labour of love and hard work preparatory to the meeting of the British Association at Aberdeen, at which I shall have to hold forth on the subject." (*Geikie* II 251).

To: Falconer
From: Birnie
Date: August 1859
DDW 71/859/7

I shall not attempt to describe but leave yourself to picture the joy
which your last letter diffused through the hearts of the Heads of the
family at the Manse. We are deeply thankful that through your great
kindness our second boy has now the fair prospect of a good start in the
race of life. May God reward you for the helping hand you have given
him. After weighing the matter I find that, with justice to the other members of the family, I could not afford to keep him three years at Addis-
combe, and therefore must let him make his way in the infantry. He will
be in good hands with Mr Morrison who has just given up the Elgin
Academy & set up an establishment for training boys for further situa-
tions, and as he will [be] anxious early to establish the reputation of his
seminary, Mr M will make a point of doing all [   ] those of his pupils
who earn the position of [   ].

On Wednesday week I met Sir Roderick Murchison & Mr Ramsay at
Forres &, like the hunted fox, we looked into every hole along the north
side of the Pluscarden Hill—from Blenie to Alves;—closely examined
the passage of the red beds of Newton Quarry into the <u>white</u> of the
Knock; then inspected the *Holoptychian* bearing strata of the Hospital
and Laverockloch quarries;—then to the Reptilian beds of Findrassie
and Spynie back to the *Holoptychian* quarry of Bishopmill. Sir Roderick
from an attack of diarhea was unable to go out with us next two days;
but I lead Prof Ramsay over the ground on which Sir R went last year
viz at Lossiemouth, the Moniack Hill past Birnie to Elgin. Next day
we took the Coast section from Burghead to Covesea, while Sir R
was able again narrowly to inspect the strike & dip of the reptilian &
Holoptychian beds where they come nearest to each other at Findrassie and Bishopmill. On the further day we were all on the Stotfield
Skerries etc.

Whatever palaeontologists may say, the result is that the Reptiles will
be considered as firmly fixed in the Old Red as they were last year. The
stratigraphical evidence is wholly in favour of this view, save only actual
contact.

But of this you will hear more.

To: Gordon
From: Jermyn Street
Dated: 6 October 1859
DDW 71/859/35

On my arrival here I learn that Sir C Lyell has come to the conclusion that Mr. C Moore[95] (for <u>he</u> it is who has done the thing) is right, & that the Reptilian Sandstone is Keuper i e Upper Trias.

All this must of course depend <u>exclusively</u> upon the Linksfield quarry which I was too unwell to revisit, which Ramsay did not see, & of which you made lightly.

How Lyell has made out the <u>clear</u> distinction between the two Sandstones is more than I can comprehend—& how he is to account for the case observed by Ramsay and yourself it is difficult to imagine.

The palaeontologists are in delight, & if it be <u>really proved</u>, I shall rejoice also;—but I cannot help thinking that wishes & the usual order of Nature have had more to do with this result than any stratigraphical <u>proofs</u>.

Let me know what you say about it & if you have been converted?

As all your friends including myself desired to see you honoured in your old Alma Mater I wrote first to Lord Aberdeen, but then found you were of the Marischal College & that Chancellors had nothing to do with LLD's.

Thereon Prof Nicol who said he had before thought of it told me on my applying to him that he could manage the affair.

To: Murchison
From: Birnie
Dated: October 1859[96]
DDW 71/859/42

I am much gratified by the assurance which your note, just received, gives me that you have returned safe and well to London, after your prolonged and laborious work in Scotland;—and at the same time I beg to assure you that I still am, after all that I have seen and heard, <u>un</u>converted to the belief that the Elgin fossil Reptiles belong to the Trias. It is, in my opinion, as difficult as ever to separate them from the white & yellow *Holoptychius* bearing beds of the Bishopmill or inland ridge of Sandstone.

It probably may be that Mr Moore is correct in calling the Linksfield beds Lias running into Trias; but this does not remove the difficulty—all who

---

[95]Charles Moore (1815–1881), geologist, left his fossil collection to the Royal Literary and Scientific Institution, Bath. See "On the so-called Wealden Beds at Linksfield, and the Reptiliferous Sandstones of Elgin" *QJGS*, 16, (1860): 445.

[96]This letter was written between the 6th and the 12th of October. On the 6th of October Murchison writes to Gordon "Let me know what you say about it and if you have been converted." In this letter Gordon replies that he is "unconverted."

have recently seen these beds—Sir Cha[s] Lyell, Prof[r] Harkness[97] and M[r] Symmonds[98] allow that they are not only <u>not</u> in situ but that they actually <u>rest</u> upon a bed of Till or Boulder Clay of the same nature as that which <u>covers</u> them. And on Tuesday week, D[r] Ross[99] having sent a person to clear away part of the Till from the underlying Cornstone, I showed and convinced Prof Harkness & M[r] Symmonds further that the surface of Cornstone had been powerfully affected by glacial action. Sir Cha[s] Lyell was not with us at the time, having gone into Elgin but he suggested, as the more probable explanation (than the calling in of an iceberg) of this singular position, that there might have been an old shore on which some Till had been laid, and that then by some undermining forces, the beds of Linksfield had slid down from some high ground supposed to be then existing in the neighbourhood to then present site. He extended some similar cases he had seen on the Continent of Europe. Be this as it may no passage can be traced between the Linksfield Liassic beds & the underlying cornstone or between the Liassic beds and the Reptiliferous Strata of Findrassie and Spynie.

Owing to the state of the tide none of these gentlemen saw the patch of rock, containing shells, which Prof[r] Ramsay and I found at Stotfield. The flinty agatey rock at Stotfield seemed to surprise them much. Prof[r] Harkness, so far as I could gather from his observations saw no reason, on this his second visit, to alter the opinion he had given at Aberdeen viz that these Reptiliferous beds were either the upper beds of the Old Red or the lower of the Carboniferous Series. The other two, Sir Cha[s] Lyell & M[r] Symmonds seemed strongly to believe that they would <u>ultimately</u> be found to be Triassic; but still regarded it, and think all must do, as yet who have seen the ground, still an open question, and very properly suggested several points of observation which should claim the future attention of the local geologists in order to clear up the mystery. M[r] Symmonds on several occasions, particularly on the sea coast ridge said the general aspect of the rocks forcibly reminded him of the Trias in some parts of England.

Here take in the notice of M[r] Duff's rib sent to Prof Owen.

You recollect that in the section which he gave of this district at Aberdeen Prof Harkness shewed a fault as occurring at the mouth, or south end of the Bishopmill quarry—and which appearance you & Prof[r] Ramsay

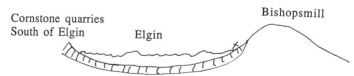

Cornstone quarries
South of Elgin   Elgin   Bishopsmill

---

[97]Professor R. Harkness. In his Aberdeen paper entitled "On the Yellow Sandstones of Elgin and Lossiemouth", Harkness said: "The rocks are to a considerable extent marked by debris; but whenever these are apparent, they manifest no traces of faults of such an extent as would disconnect the *Holoptychius* yielding strata from the reptilian beds which occur in this portion of Moray." A re-worked account of the same material appeared as "On the Reptiliferous Rocks and Footprint Bearing Strata of the North-East of Scotland", *QJGS*, 20, (1864): 429–443.

[98]The Rev. W. Symonds of Pendock Rectory, Tewksbury.

[99]Dr. Ross of Pitcaline, friend of James Joass.

saw. Taking this in connection with the dip of the Cornstone at Linksfield and a very indistinctly stratified piece of Cornstone on the Lossie, halfway between Linksfield & the Hospital quarry, Prof^r Harkness is led to think that the Cornstone, on which the town of Elgin stands, is troughed, thus He made a fuller and more distinct section of which I am sure he would readily furnish you a copy—There are some difficulties which this suggests to my mind but as it does not by any means affect the question of the age of the Reptiliferous beds, I shall not here enter upon them.

Another circumstance has come to my knowledge which may have some weight on the question. On the last day that Sir C Lyell was at Elgin Mr Duff showed us a letter he had seven years ago from Prof^r Owen. The following is a copy <u>of a copy</u> of it. "College of Surgeons January 14, 1852 Dear Sir. I have only time to thank you by this post for the *Leptopleuronic* rib which safely arrived and is identical in microscopic structure and in form with those of the fossil skeleton. Of the purely lacustrine affinities of that skeleton I have no doubt, and others are coming round with more or less ardour to that opinion. But what seems of more consequence is to have every doubt cleared up as far as possible relative to the age of the stratum. What may be the exact locality of the fossil which reached me this morning? I shall have a careful drawing made of it & return it to you by post. Very truly yours etc (signed) P. Owen."—Unfortunately this rib is lost. At least it never reached Mr Duff. It was got at Linksfield and the skeleton to which Prof. Owen refers I have little doubt was the Spynie Reptile *Telerpeton Elgin* (his *Leptopleuron lacertium*). All then that will ever likely be seen of this rib will be the drawing of it which I hope was made at the time by Prof Owen.

Among the specimens returned to the Elgin Museum from Aberdeen, the bone which Mr. Martin gave to you has been found. It will be carefully transmitted to Jermyn Street by the first opportunity.[100]

To: Lyell
From: Murchison
Date: 12 October 1859
ALS: BD25.L

On my arrival here I was informed by Huxley of your having decidedly adopted the opinion that the Reptilian Sandstones were of Triassic age & on <u>stratigraphical evidence</u>.

I also found that Falconer & others had been talking much of this determination.

---

[100]The following was crossed out at this point: "Upon the whole question as to the age of the Reptiliferous beds of Findrassie, Spynie, Lossiemouth, and Cummingston is open as ever the stratigraphical argument remaining <u>untouched</u>". The following was written across lines near the drawing: "the question still remains open the stratigraphical argument remaining untouched." Having been undecided while composing the draft, Gordon included these lines in the letter to Murchison as Murchison refers to them in the letter to Lyell on the 12th of October.

Now as such a determination involves of necessity a serious charge upon the capacity for field observation not only of myself but of Ramsay, Harkness, G. Gordon & others who simply say that they can observe no stratigraphical proof of the disconnection of the one sandstone from the other, so I did hope that you could have made the earliest communication to me if the grounds (physical data of course) on which you had arrived at this conclusion—a conclusion I am most willing to adopt though I do not like to be shown up as having observed inaccurately.

I find by letters from Harkness & George Gordon that they both consider (after your Survey) the 'stratigraphical case to be untouched' & that no new proofs were observed for I hold that the Linksfield case is worthless.

Let me know then if you have anything else except the strong disposition which I have in common with yourself to believe that the deposits may prove to be of Triass age.

To: Lyell
From: Murchison
Date: 19 October 1859
ALS: BD25.L

I thank you for your long, clear & well reasoned letter of the 16th on the reptile sandstones of Elgin which I had on coming to Town yesterday from Tunbridge Wells.

It would ill become me to set myself in opposition to inferences which are quite in accordance with my general views & wishes. All that I can hope you should do, is to admit, that a person viewing the subject as I did in an ascending order which seemed un-broken would actually see rather greater agreements between the two sets of sandstone than you seem to have noticed.

Did you, for example, visit the <u>new</u> quarries visit the new quarries[101] of *Holoptychius* <u>yellow</u> sandstone in the woods above New Spynie to the West of the Reptile Findrassie quarries at a place called <u>Laveroch Loch</u>?[102]

---

[101]The phrase "visit the new quarries" was inadvertently repeated in the letter.

[102]Laverock Loch, two miles NNW of Elgin. In this paragraph, Murchison is referring to a day recorded in notebook M/N 135 where his entry reads as follows: "To the S of New Spynie—i e over the higher wood & in an adjacent portion of it near to a local depression called Laveroch Loch, quarries have been opened since I was here & which having afforded the scales of *Holoptychius* are of great interest in determining the relations of the Fishy Old Red of the Elgin Ridge to the reptiliferous strata of its Eastern End—i e at Findrassie & Spynie [Quarry]. Those *Holoptychian* sandstones of Laveroch Loch were found to be [entirely] hard & siliceous on their surfaces—(like those on the summit of the Knock of Alves) & like those of Findrassie & Lossiemouth. But on quarrying there, they afforded the finest examples of soft sectile yellow sandstone with some white beds. These beds dip to NNW or N by W at a slight inclination like all the other masses of the same ridge.

Proceeding on the strike to ENE you next come to the celebrated Quarries of Findrassies— which have furnished reptilian remains & in them we have the same mineral characters—& the same strike & inclination. The [succession] there laid open in [the] [sets] of adjacent quarries is as follows:

If you did, I am sure you must have been struck with the identity in character, strike angle & direction of dip of the two sandstones. It was this additional evidence which most affected my reasoning however for the one with fishes seemed then as far as I could observe almost an absolute continuation of the other—& not distant.

Again, surely there are yellow sandstones at the Masons Haugh & footprint quarries of the [Coast] ridge which it is scarcely possible to distinguish from those of the *Holoptychius* Elgin ridge?

I shall get up specimens to show them at all costs for exhibition in our Museum, for I wish to bring the two sandstones into fashion for our building purposes in the Metropolis.

As to the supposed fault of Harkness, as [deduced] from the appearances at the S end of the Bishop Mill Quarry you must not suppose that Ramsay & myself were so blind as not to observe them.

I specially called my companions attention to it but he viewed it as one of those local breaks so common on escarpments & not a persistent fault, for we could not trace it to the E & W. This I have explained to Harkness;

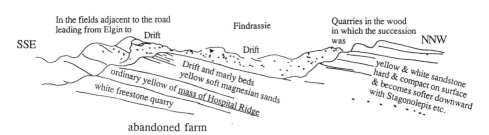

abandoned farm

Now the strata in which *Stagonolepis* occurs are important in two points of view. 1st. They strike to the ENE with the *Telerpeton* quarries in Spynie Hill & both of these dip to the NNW. 2nd. They offer at this [ ] [proximity] the same descending succession which is seen at Lossiemouth—i e red & yellow soft marly & sandy mottled marly yellow beds which connect the reptilian beds with the mass of Elgin Ridge in which *Holoptychii* occur—whether at the Bishop Mill quarries near Elgin or its southern side where [ ] yellow sandstones also dip to the NNW & pass directly under the Findrassie & New Spynie strata above mentioned—as at the Hospital & Newton Quarries. In all of these scales of Old Red Fish occur.

From these data it follows that there is not a scintilla of evidence to show that the reptile beds can be of Triassic age—On the contrary the geological inferences as derived from similarity of strike & dip as well as from the [ ] *Holoptychian* beds being lithologically [indistinguishable] from those with reptiles lead me to conclude with still greater decision than I [ ] last year that the reptile beds are a regular & [correlative] sequence of *Holoptychii* beds.

Last year I was not aware that *Holoptychian* scales had been found in the Hospital quarries though I had seen them in the Bishop's Mill & Newton quarries both distant on the Elgin ridge. And now that the important addition has been made of detecting said Old Red Fishes in the new quarries on the very strike of the Reptile beds & within ¾ mile of them—beds so situated that if inferior they must be very slightly inferior to the Reptile beds I have no hesitation in believing either that the latter do really constitute the uppermost band in the Old Red series, or some band of transition upwards into the next overlying nature Palaeozoic deposits as I have stated in my memoir." Lyell would have immediately seen the irrelevance of the similarity of strike and dip at Laveroch Loch with Spynie Wood.

also in sending me a columnar section of his hypothetical view & section has assured me, that this section makes no change <u>whatever</u> in his opinion as expressed at Aberdeen.

But even if I were to grant this fault to be true, it would not bring <u>Linksfield</u> (which lies far to the East) into a section due South; for the first cornstone met with in any sectional line southwards through Elgin shows all the cornstones dipping <u>NNW</u>.

Further I am convinced that the Cornstone of Elgin proper is absolutely that of Cothall on the Findhorn i.e. a limestone quite <u>subordinate</u> to & intercalated in <u>fish</u> beds.

The Cornstone of the Castle of Spynie & of the [Coast] ridge are Upper in position & clearly different in Mineral character. Now, if you or Mr. C. Moore can show that the Cornstone of <u>Linksfield</u> is equal to that of Lossiemouth & the shore, you will I think have made your case much more rational.

For my own part as I think I told you before, I was seriously ill & in bed two days at Elgin whilst Ramsay & Gordon were exploring & I did not visit the Linksfield beds which I saw cursorily only in 1858, & long ago in 1827. But my fixed impression was, that that cornstone was the Middle or Cothall cornstone & not the supra-reptilian cornstone.

Pray pay attention to this process in my reasoning & give me credit at all events for not seeing how it was <u>for me</u> to separate these sandstones.

I have a capital collection of Liassic fossils from . . . .[103] But then in looking at the occurrence of these fossils at various points & as it seemed to me in <u>various</u> & <u>different</u> strata—(sometimes in a limestone sometimes in a sandstone as far as we could judge from debris) I looked to those as proof of a <u>transgressive</u> deposition.

Above all I am struck with the fact, that whilst in Brora & all along the coast of Rossshire the Lias & oolites were entirely unconformable to the Old Red—the yellow & white sandstones with Reptiles were perfectly parallel to the yellow & white sandstones with *Holoptychius*.

I am sure that it is not your wish not fairly to admit to difficulties of the case & at all events to allow that I endeavored to act fairly in so stating it, as to allow the probability of a balance of opinion on that side to which all my previous views were opposed.

Yours very truly
Rod I Murchison

P.S. Putting aside the upper Elgin Sandstones I am happy to tell you that my Highland Reform Bill has received a great accession of strength by Harkness'. He has given in to several details which he sends to me for publication with my own observations & those of Ramsay which do not leave the smallest doubt as to the superposition of the eastern or Upper Gneiss to the Lower Silurian fossil band.

---

[103]Murchison could not remember the name of the site where the Liassic fossils were collected.

When we meet I have much to tell you of what I saw in the Kincardine & Forfar tracts where in the Grampian Valleys my Silurians are little metamorphosed & where the Lower Old Red with [   ] flags is on the <u>west</u> side <u>intercalated</u> in great conglomerates.

To: Gordon
From: Robt. McAndrew[104]
Dated: 29 October 1859
DDW 71/859/41

I yesterday gave the Lossiemouth fossils to Huxley who is much pleased with that containing teeth of both upper & lower jaws. Our friend Mr. P. Duff made a mistake in wishing his specimen to be shown both to Huxley & Owen. After calling at the Museum twice without being able to meet the latter, I left the fossil in charge of S.P. Woodward[105] with instruction to shew it to Owen & afterwards forward it to Huxley, but he (Huxley) hearing that it was to be in Owen's hands would have nothing to do with it, & said he would write Woodward to return it to Mr. Duff—so the matter rests—I of course do not mix myself up in these squabbles—You will probably have an opportunity of explaining to Mr. Duff, the state of the case.

We reached home on Tuesday last in frightful weather, after being nearly snowed up at Carnousie. I am happy to say however that my wife is none the worse, but on the contrary her health has benefited by the change—She unites in best regards to Mrs. Gordon & your family.

To: Murchison
From: Birnie
Dated: 22 November 1859
LGS/M/G10/3[106]

Yesterday I received the gratifying communication from Marischal College, Aberdeen, that the Senatus had unanimously conferred upon me the Honorary degree of L.L.D.

Without delay I must acknowledge with gratitude the kind and influential part which you have taken in obtaining for me this unsolicited, but I fear unmerited, distinction.

---

[104]Robert McAndrew (1802–73), merchant, yachtsman and naturalist. FRS 1853. Like George Gordon, Robert McAndrew had been an undergraduate at Marischal College.

[105]Samuel Pickworth Woodward (1821–65), Assistant in the Department of Geology and Mineralogy, British Museum (1845–65), was the author of *A Manual of the Mollusca; or, a Rudimentary Treatise of Recent and Fossil Shells* in three parts (1851–56). In his letter to Gordon of 29 October 1859, Huxley said: "For very good reasons which you will readily understand without my troubling you with them at length, I immediately declined having anything to do with the specimen and I wrote to Mr. Woodward in whose charge it is, requesting him to return it directly to Mr. Duff. I also wrote to the latter gentleman to tell him what I had done."

[106]The draft of this letter is DDW 71/859/47.

To: Murchison
From: Birnie
Dated: 22 November 1859
DDW 71/859/47[107]

Yesterday I received the gratifying communication from Marischal College Aberdeen that the Senatus had unanimously conferred upon me the honorary degree of L.L.D.

Without delay I must acknowledge with gratitude the kind & influential part which you have taken in obtaining for me this unsolicited but I fear unmerited distinction.

To: Gordon
From: [Jermyn Street]
Dated: 1 December 1859
DDW 71/859/52

My dear D̲ͬ Gordon,

The announcement of your <u>Doctorate</u> both from Prof Nicol & yourself, gratified me much. It is one of the good results of the gathering at Aberdeen, which met to advance Science in the North and your old Collegiae acted in the right principle of *"Premiando Incitat."*

So I dare say you will become more vigourous than ever, & will furnish us with some more reptiles.

We have been much startled by the apparition of Darwin's book on the "Origin of Species". Huxley is quite a believer in this ultra-Lamarckian theory of "<u>Natural Selection</u>" & the "<u>Struggle for Existence</u>", by which one race has passed according to general laws, into another; & so we all descend from a Monad up thence as from inferior types. <u>I am decidedly hostile to the whole thing</u>, & am as firmly a believer as ever in Successive <u>Creations</u>. By the bye that view of Creation is not in Darwin's book & I do not therefore think it will suit <u>Your book</u>.

The most satisfactory addendum to my Campaign in the N west Highlands, was that the subsequent visit of Prof Harkness completely confirmed (perhaps I did not require that) & even contributed further illustrations of the truth of my postulates. I will send you the abstract of my last. This great & fundamental reform of the Highland or N Scottish rocks is quite apart from our interesting slice of land which as yet belongs to 'No man'—i.e. Geologist—

With kind regards to all your family

---

[107]The final version of this letter is LGS/M/G10/3. See previous letter.

To: Gordon
From: [Jermyn Street]
Dated: 14 May 1861
DDW 71/861/3

Thanks for your letter of the 10th.

On consulting Huxley & others I find that it would not be worth our while to give more than £40 for Mr. Patrick Duff's collection,[108] irrespective of the Cabinet which would be of no use to us.

I do not think his Executors will get more—probably not as much by breaking it up. Supposing the two crack things to fetch £20—the remainder (chiefly fragments) do not seem to be worth more than £20.

If this sum of £40 exclusive of the Cabinet be entertained you could be so kind as to let me know.

To: Murchison
From: Gordon
Dated: 27 May 1861
DDW 71/861/3[109]

Upon the receipt of your last note I communicated its purport to Mr Duff's Executors.

They have now informed me that they regret they cannot close with your offer, more from a regard to the expressed feelings of Mr Duff's sister, who as her brother did, places an undue value on the collection, than from any expectation of their getting more for it from any other quarter. I believe they themselves agree with what Dr Taylor, Mr Martin & myself have said that the offer from Jermyn Street is a very [fair] one but they feel that they cannot at present accept it.[110]

To: Gordon
From: Jermyn Street
Dated: 10 June 1862
DDW 71/862/12

I take great shame to myself for not having sooner replied to your letter of the 7th May.

At least I fear I have omitted to reply.

---

[108]As already noted, Murchison had first seen Duff's collection in 1840 and then again in 1858 (M/N 134, folio 52).

[109]This draft was written on the letter (DDW 71/861/3) received from Murchison.

[110]The following was crossed out and replaced by the last sentence. "They have my own and Mr. Martin's & Dr. Taylor's opinion that the School of Mines in Jermyn Street have made them a fair offer and I believe they think so themselves."

If however the choice specimens of Mr Patrick Duff to which you allude are still to be had, I beg to say that although there is no money forthcoming at this moment in our account I am willing to purchase them for this Museum on my own risk at any reasonable price which you may think fair & equitable.

I will trust it all to you & hoping that the things are not sold.

As chairman of a very, very [   ] class of the Exhibition No 1 I have been incessantly occupied & your letter was mislaid.[111] [112]

To: Murchison
From: Birnie
Dated: [July 1862][113]
DDW 71/862/12

I yesterday by chance met one of the late Mr Patrick Duff's executors, who informed me that his Cabinet of fossils would be brought to the Museum at Elgin on the end of this month. His executors much regretted that owing to the undue estimation that M^r D's sister had been taught to form of the collection [of] specimens, they were unable to accept of the liberal offer formerly made by you.

I have already stated that, taken as a whole the collection is not by any means so complete or select as I once regarded it, yet the *Telerpeton* & the casts of the caudal vertebra of the *Stagonolepis* (both well known to Professor Huxley) are certainly valuable.

Should you still desire to have them for the Jermyn Street Museum be so good as [to] drop me a note limiting me to a sum and I shall endeavor to purchase them for you at the sale as much under it as the competition (if any) will admit.

To: Murchison
From: Birnie
Dated: 30 July 1862
DDW 71/862/12

Aware of the interest you always take in the geology of the North of Scotland[114] and particularly of Ross-Shire, I have the pleasure of sending you

---

[111]This last paragraph appears after the signature. Geikie quotes a letter to him from Murchison in which Murchison said: "I have had, besides, very hard work as chairman of Class I of the Exhibition. With 2000 exhibitors in my class, it has been no small difficulty to adjudicate with fairness some 300 or 400 medals" *Geikie* II 255.

[112]The following in Gordon's hand appears in the margin of Murchison's letter: "Sir Rod Murchison etc. about Mr Duff's cabinets & 2 letters to do with Rosshire trails etc."

[113]This letter is attributed to 1862 because it is part of an intelligible sequence.

[114]Of those letters which have survived, this one marks the re-commencement of Gordon's relationship with Murchison as he, Gordon, more and more turns his attention to the Black Isle. Huxley acknowledged the letter on 2 August 1862. "In Sir Roderick's absence (he is in

a photograph & sketch (by Mr Joass of Eddertown[115]) of some footprints or track of an annelid or Crustacean (if not of a Reptile?)

The slab 1 foot 6 in by 1 foot 2 in was found some weeks ago just turned out of a quarry a mile to the east of Portmahomack.[116] It was found by Mr Campbell minister of Tarbat who was in company with Mr Joass. To this latter gentleman I am indebted for the illustrations I now send you. Last year I examined the full section from Nigg to Tarbatness lighthouse. At many points in the eastern portion of it I was reminded of our upper or reptilian beds on this side of the firth. We shall be anxious to hear to what class the creature that made their marks probably belonged. I have not yet seen the slab but am promised to have it in my possession soon.

Mr Patr. Duff's Cabinet is now placed <u>for sale</u> in the Elgin Museum where it is more likely to come under the eye of a purchaser. I hear that the Executives are to sell it <u>as a whole</u> & thus save geological [unity] as to the price have been been making inquiries.

I could not venture to offer more than the handsome sum you stated last year.

To: Murchison
From: Birnie
Dated: 2 August 1862
DDW 71/862/17

I have just recd the enclosed from Mr Duff's Executors, & also obtained their permission to submit it you.

I shall be guided by your reply, with which I trust you will favor me by return of post.[117]

To: Stables
From: Joass
Dated: 8 November 1862
DDW 71/862/38[118]

You may believe that I was very much delighted to hear again from you after a silence which I can now understand but which led me at first to

---

Bohemia & will probably be absent until the middle of October) your letter of 30 July with the enclosure from Mr. Joass has been put into my hands. I wish I could give you an opinion worth having on the tracks, but beyond feeling pretty sure that they were not made by any vertebrate animal, and that they were made by some Armeloid creature, I have not much to say about them."

[115]The Rev. James Maxwell Joass, pronounced (Joss), at least in the North.

[116]The assistant curator of the Elgin Museum informs us that there is just one slab containing footprints from Tarbet Ness (Elgin Museum reference number 1863.3) in the museum.

[117]This letter, too, was acknowledged by Huxley, Murchison being away for the summer. See *Huxley at Work* p. 122.

[118]This letter is in the Gordon archive.

think that my letters had gone on lost (Rossre) or that I had seriously offended you, which notion troubled me much.

I should have replied sooner but wanted to tell you that I had been over the section from Nigg ferry to Shandwick & think that if the weather prove favourable you would be much pleased if you carry out your intention of coming by boat—I hope that should you & Dr. Gordon come you will ask me to meet you anywhere on the route or at Cromarty & will arrange to come here if your plans permit—as I have some fine ichnites to exhibit.

At the base of the North Sutor the sandstones[a] are intersected by 6 limestone & shaley bands[b], containing scales and fucoids. The beds dip very abruptly WNW—then comes a large mass of Jaspery Conglomerate[c] followed by a quartzite & some ironstone & flaggy Silurian rocks[d], containing caves. About 1 $\frac{1}{2}$ miles eastward the flags become much contorted & broken up by Felspathic & quartz veins[e] at which point masses of a black micaceous rock intruded by quartz appear[f]. The dip after this is eastward, curving round to North as you approach Shandwick. About 1$\frac{1}{2}$ miles from which the Conglomerate reappears[c2]—while blocks of limestone occur on the beach[b2]—then follows a patch of limestone in situ[b3] & then the sandstone[a2]. The beds being pretty regularly stratified & dipping apparently NNE at a low angle & occur nearly level in some places—

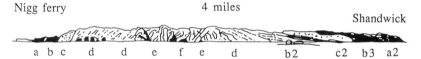

Nigg ferry   4 miles

                                Shandwick

a b c   d   d   e   f   e   d   b2   c2 b3 a2

I am very glad to hear that Mrs Stables and the children are well, and that Miss Mary is practicing on the Carpenter—I hope she takes the full value of her attentions out of him in the shape of wheelbarrows, spade handles etc. with bricks enough to build a Babel with.

To [recur] to the rocks, do you think that gold might be found in the Quartz veins connected with the altered sandstones of the section referred to.

Mr. Campbell reports a new & larger track at Tarbat. Very indistinct—I shall be there first week and shall let you know if it turns out interesting.

Weather bad today. Fresh snow all round this morning & frost below.

Hoping to hear again from you if you think of coming across some calm day some of which I hope we may yet have ere the weather breaks into winter.

To: Murchison
From: Birnie
Dated: 21 November 1862
DDW 71/862/40

At the end of July last, while not aware of your absence from London, I sent you a photograph of a slab (shewing trails of different animals) that was discovered in a quarry near Portmahomack, by the Revd. George Campbell of Tarbat. I am glad to see, from The Times report of the last meeting of the Geographical Society, that you have returned from your continental tour, and I trust invigorated thereby for the labours of the coming winter. No one I am persuaded will rejoice more than yourself at the progress geological discovery is making in Ross-Shire your native country. It therefore gives me pleasure to inform you that my friends Mr Joass of Edderton, and Mr Campbell have found slabs, all from a quarry near Portmahomack, that shew the trails or tracks of I think five (at least) different animals.

Professor Huxley kindly acknowledged the receipt at Jermyn St. of my note & the photograph of the slab, which I shall call 'A', and intimated his opinion that the trails on it were not made by any vertebrate animal. This view has been confirmed by Mr Page, who saw the slab (A) since its arrival at the Elgin Museum. This slab exhibits two different tracks viz No. 1 which is one and a half inches broad, and occur over the whole length (8 inches) of the stone. It consists of four parallel lines the central two being continuous (with length) the grooves probably made by the tips of the animal's tail. The lines at the side are dotted and seem to be the footprints.

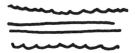

The next form of markings (No 2) which also run across the whole surface of the slab, has but two lines which are dotted, and only half an inch apart.

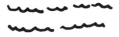

The third (No. 3) set of markings were discovered by Mr. Joass and are seen in the annexed photograph, particularly in the lower figure, and at the upper corner, left-side. I also append a Joass sketch of them, cut from Mr Joass letter, 1 Sept, which, he says, "shews about the true size of these dotted lines which are ranged diagonally to the line of section, and which are continued from one end of the stone to the other. In reference to these markings (No. 3) Mr Joass further adds, that he set 7 or 8 crabs arunning sideways on damp sand; and after careful examination of these tracks, he was all but convinced that these (No. 3) Tarbat tracks were formed by Crustaceans of a sort very like our smaller crabs. "The number and diagonal character of the punctures correspond, and one [litter] nicely [streak] over as [   ][   ][   ] that the 5 & 6 holes were sometimes very close to each other and occasionally are together, both on the sand and in the stone—

they being the double puncture made by the front claws more or less opened.

But what I hold as the most important of the whole slabs which these Scotsmen have discovered at Portmahomack is slab "B" which I have just despatched [   ] from Aberdeen, to your address in Jermyn Street. On it there are two well defined sets of markings viz No. 4 & No. 5. No 4 consists of two parallel lines which are both fringed on the same side. In the rough sketch, which I have sent as a sort of guide to the different parts of the slab, the fringed lines appear at the places marked 'No. 4' and I think at other portions of the stone.—It is however to the other and larger set of markings on this slab that I [earnestly] by your closest attention, and the similar attention of your ichnological friends. To see these large markings (No. 5) the slab gives the strongest indications of being reptilian footprints. These [missing line?] side of the Moray frith; but the [bilobular] outline of several of them as at "B" on the rough sketch, is remarkable. It is also to be observed, particularly in the [course] of the red line, that the smaller of these lobes is always in the inside of the footprints If the sand or clay was very soft or watery, when the animal passed over the hollows left by the feet would be filled up with water, which would wash away the sharp angles of the impression.

But even if ever granted [   ] that these were the footprints of a reptile, the question would be immediately asked what is the age of the Tarbatness Sandstones?—as it has been put in reference to our Elgin reptiliferous beds.

In September 1861 I walked from Shandwick to Geanies, & then from Portmahomack round by the Lighthouse back to Geanies, thus completing a hurried inspection of the peninsula. In many places, particularly between Portmahomack & the Lighthouse, I saw a marked resemblance to the rocks in our Burghead and Covesea range. Mr. Stables, who was with me, remarked at the time, that that ground would probably prove the key to the solution of the questa vexata, the age of our upper Elgin beds. From Shandwick down to and round the Lighthouse, at Tarbatness, I could see no break or unconformity, no great transition of lithologic character, or such difference of dip or strike in the strata of the whole range as to raise any doubt of the rocks of the peninsula belonging to the same formation. And a finer and more complete section I have never [   ] It is never wholly obliterated at any place; but may be read off continuously, from Shandwick to Portmahomack. But still the necessity remains for a further search & further finding of a holoptychius scale, or some fossil from those beds, before the question ceases to be an open question, which with all due deference, I [maintain] still to hold it. Triassic or Devonian?—And now that these ichnites have been found at Portmahomack, the interest of the [   ] begin to rise with geologists. I propose therefore, as soon as the winter months are past, to pay it another visit. Meantime no fitter person could be on the ground than Mr Joass, who has just written me that still further traces of animal life have been got from the same quarry. Notwithstanding the advance of the season, he lately took a boat and examined the coast from Nigg to Shandwick. The accompanying sketches shews his accuracy as an observer & his capability & [ ] as a draughtsman.

The rocks at Nigg, nearest Cromarty, and marked d′ are a good base or starting point for a searching examination, as they contain undoubted Old red fossils—osteolepis & fucoids. Still better [ ] would be at d & c² both parts nearer Shandwick, if they shall be found to be [ ], as I suspect [ ] d′. The bed e which Mr Joass calls jaspery conglomerate, will possibly be a brecciated quartzite.

I fear I have [imposed] too long on your time and patience by this communication. I shall however [transmit] to your notice from time to time of any future discoveries [from] my work.

To: Gordon
From: Murchison
Dated: 24 November 1862
DDW 71/862/41A

A thousand thanks for your excellent Dispatch which after the inspection of the Slab shall be presented to the Geological Society by me & properly brought out as a Memoir to serve in the elucidation of the *questio vexata*.

Huxley looking at your diagrams has no doubt that the foot prints are those of a Reptile.

On the other hand Ramsay & myself are delighted, for we now feel pretty sure (having walked nearly all the section from Shandwick to Tarbat Ness & from the [lighthouse] to Port Mahomack) that if not [persistently] all Upper Old Red it is at all events of Carboniferous age or a Paleozoic rock & so likened as to the Old Red that Trias or Mesozoic age is out of the question.

You have put me in good spirits—Field work for ever & perhaps the proud palaeontologists will be obliged to bow to stratigraphical sequence.

p.s. How will you, or can you, then connect the yellow beds of P. Mahomack to the yellow beds of <u>Dornoch</u>? which we classed with the Old Red Flags.

To: Murchison
From: Joass
Dated: 15 December 1862
LGS/M/J5/1

Encouraged by your last kind letter I have been carrying on my search among the Tarbat Ichnite beds in company with my friend the Rev. Mr.

Campbell of that Parish and, thinking you may be interested in the discoveries made there I take the liberty of enclosing a sketch of some of the prints last found. They occur on and also a little above the bed on which the first crustacean-like traces were discovered but considerably to the westward. Searching along the strike toward the point of Tarbat Ness I found similar tracks near the Lighthouse.

I may also mention that during a recent examination of a limestone belt at Geanies, interstratified with clay shale and beds of coarse nodular limestone I found several of the limestone flags thickly covered by scales of the osteolepis & fucoids, besides a few small Coprolites & part of a plate of the *Coccosteus* or pterichthys.

The Ichnite beds occur about half a mile to the North East of Portmahomack, about highwater mark, the dip being a little West of magnetic North.

To: Gordon
From: 31 Regency Square, Brighton
Dated: 19 December 1862
DDW 71/862/49

A thousand thanks for your most gratifying & acceptable letter of the 17th with the graphic & clever enclosure from Mr. Joass.

I have written at once to this capital & zealous fossil-hunter & sketcher & have backed up your request & have urged him to write a Memoir for our Society volumes.

It is indeed [splendid] that the geologists happily so called should have this sweet revenge on the palaeontologists.

I have no manner of doubt that all these yellow sandstones are Palaeozoic.

To affirm that they ought to be defined as Devonian or Old Red would be too much—until they find O R fish scales with them or above them.

They may possibly if not probably represent the Lower Carboniferous of the S of Scotland which is charged with Reptilia—2 or 3 new sources having been recently discovered.

I am here for a week with my wife who has been out of health & before the New Year shall be again in London.

To: Gordon
From: Brighton
Dated: 21 December 1862
DDW 71/862/51

Your letter with the fragment containing Fish Scales from Geanies has just arrived & will be duly reported on by ichthyologic authority when I go to Town.

In the meantime I must inform you that there is nothing (as far as my recollection serves me) <u>new</u> in this find of Mr. Joass.

If Sedgwick & Self in 1827 did not find fishes we alluded to them in the bituminous schists of <u>Geanies</u>.[119] And in subsequent writings I have always considered the Schists of Geanies & Strathpeffer to be quasi = to some member of the Caithness Series.

The [true thing   ] to be done is to work out precisely the succession from Geanies to the Light House.

It is, if I mistake not, just the case of Dunnet & Ross Head over again, in both of which the sandstones are superior to the Caithness Schists & constitute in my map be No. 3 of the Old Red Series. This No 3 may prove to be the Lower Carboniferous; but as yet I hold firmly to my original suggestion, at all events until it be shown, that there is a great <u>discordance</u> between the Fish beds & the Reptilian strata.

To: The Rev. W. Symonds
From: Birnie
Dated: 31 December 1862
Lyell papers, University of Edinburgh

I had much pleasure in the receipt of your letter, and am glad to find you take so deep an interest in our northern geological discoveries.

It is but rarely that I see the Inverness Courier. Mr Stables of Cawdor has sent me a copy of the number that contains your letter.[120] It is forwarded to you by this post.

I am sorry to think that the old familiar term "Old Red" is likely soon to be expelled from our geological vocabulary, and that this once well known formation is about to be lost—one half in the Silurian, the other in the Carboniferous.—Is not this enough to call back Hugh Miller to enter his protest?—You surely ask too much of us—<u>from Geanies to Tarbatness lighthouse</u>, I grant that the theory of the Old Red age of these reptiles is not advanced.

This much, however, from my recollection of the ground, I expect he will be able to give—

The accompanying bit of scale or plate is from Geanies, and is Old Red, if the Caithness deposit is, for in the Elgin Museum the same scale or plate is to be seen from that county.

I trust that these discoveries will induce you again to visit the North of Scotland, when it will give me much pleasure to afford you any assistance in my power.

---

[119]Sedgwick and Murchison (1828), p. 150.
[120]Symond's letter referred to here and a reply by Joass appear in Appendix 7.

To: Sir Charles Lyell
From: The Rev. W. Symonds
Dated: 2 January 1863
Lyell Papers, University of Edinburgh

I enclose a copy of a note addressed by me to the Inverness Courier on the assumed Old Red age of the Elgin beds, and also a note from Mr. Gordon of Birnie which you may like to see.—I hope my letter may have the effect of stimulating the efforts of the geologists of the North of Scotland.

I would ask you to read a letter of mine addressed to The Parthenon, a London literary journal, on the subject of "Aids to Faith," the supposed <u>answer</u> to Essays & Reviews. I expect to be well abused, but I have the satisfaction of receiving some opinions corresponding with my own from others who have read my letter and the disquisitions called "Aids to Faith." I must quote a passage from a note I received this morning from one of the most accomplished naturalists with whom I am acquainted. "Mr. McCauls Essay is the most utter balderdash I ever read—A few more such "Aids" will have a very opposite effect to that intended."

Will you please remember me most kindly to Lady Lyell and begging you both to accept my good wish for the New Year.

W. S. Symonds
When <u>does</u> your book come out!

To: Gordon
From: Joass
Dated: 17 January 1863
DDW 71/863/6

The Chart, a very fine specimen of printing & so recent as to have the Invergordon Railway, not yet opened, arrived safe. I rather prematurely spoke of cutting it up into sections & folding into a case before I had your intentions concerning it. Meantime I shall copy into my sketch book on a larger scale the coast from Geanies to Tarbat & D.V. shall use it during 2 or 3 days of first week in taking notes etc. I suppose it will be enough at Geanies if I set the Laird on the scent of the Ichthyolites & get him to preserve specimens.

If the cliff section should be blind for a bit at Balone Castle I suppose the reading may be fairly taken up on the beach where the rocks are exposed at low water.

Identity of dip on to the light house should settle the sequence (?) & if a fault of recurrence happens it will but lessen the distance between Geanies and Tarbat. Should the fault result from an upheaval of the N.E. section need this make a serious difference if the dip be all right? Might a fault occur of such a nature as to result in a difference of dip beyond it & the formation be still the same I refer to

a case when the new dip is not only for a short distance but continues. What is likely to be the most prominent mark of a change of formations apart from difference of dip?

I suppose you have seen a notice of the new bird-feathers & all found in the Lithographic slate of Germany. Owen calls it the *Archeopteris Macrurus.* Chambers in his reference to the discovery (Journal [   ] 62) speaks of the peculiar markings on the old red as a feature of Modern Geology. Can Anderson's single line of tracks have been made by a bird. I return his letter with many thanks.

To: Gordon
From: Joass
Dated: 21 January 1863
DDW 71/863/7

I received your letter this morning referring to the coincidence of the Stotfield & [Causie] rocks with the line of the S. Sutor. It has been said that the water of the Dingwall frith once ran out at [Balintore], but I suspect that the Cromarty opening is older than the terraced banks of [Fearis] & Shandwick. I think that the indicated in my rough section from Nigg to Shandwick, where the rocks are split up & contorted, falling off apparently to the N.E. and [N] over Feldspathic veins and a mass of Hornblende looks like the centre of disturbance for the distance between Nigg and Tarbat point, since the dip towards the left, as you look towards the cliff, seems as much more abrupt than that to the right as the distance in the latter direction is greater than that to the N. Sutor. The upheaving force

must however have acted, though in this line, from a point to the Southward as the True dip at Nigg is not much off N.W. Could not the same action which tilted [your] beds to the N. have done the same to those on this side.

On finishing a plan of Tarbat Point, scale 3 inches to a mile & drawing a line from the beds at Portmahomack to the light house I find the dip about a point to the W of mag. North—which agrees with notes taken when last at Tarbat. I suppose this scale of ground plan lettered for reference will be large enough to accompany my sections but I find that these must be on a much larger scale as taking the highest cliff at 200 feet, which it is not, the height would only be $\frac{1}{6}$ of $\frac{3}{4}$ inch = $\frac{1}{8}$—but 200 feet would show 1 $\frac{1}{2}$ inches high in a section of 1 yard to the mile which may perhaps do—when I shall have to exhibit a slip about 6 yards long, from the west of Geanies to the Point.[121]

Since I wrote I have been consulting Jukes on stratification and think that between him & your instructions I see my way more clearly.

I wrote to Prof. H. telling him I was to try & connect Geanies with the Point by a Careful examination & promised to let him hear if I succeeded & to send him specimens from Geanies. Weather very bad here—hoping it may soon clear up when I shall at once start for Geanies.

To: Murchison
From: Birnie
Dated: 16 April 1863
DDW 71/863/38

Accompanied by Mr Stables, I have just returned from another examination of the strata from Geanies to Tarbatness. We had Mr Joass' section & illustrations before us & found them distinct & correct.[122] They will be forwarded to you in a few days in the hope that you will find

---

[121]In 1854 J. Beete Jukes set out to write an article on geology for the *Encyclopoedia Britannica* together with Professor Edward Forbes. The untimely death of Forbes delayed the article so that it appeared under the heading "Mineralogical Science." While preparing the article, Jukes felt the need of a succinct and systematic text-book to accompany his own lectures and to lead up to the study of such works as Lyell's *Principles of Geology* and *Elementary Manual*. Thus it was that Jukes published *The Student's Manual of Geology* (Edinburgh) Adam and Charles Black in 1857. Part III of that book is entitled "History of Formation of Series of Stratified Rocks."

[122]These are probably the sections and illustrations in the letter from Joass to Gordon dated 14 February 1863 and reproduced in Map 11.

them so interesting, as kindly to lay them before the Geol Society at an early meeting.[123]

A large slab with reptilian footprints from near Port Mahomack is now at Elgin. If desired I would gladly send it, as you shall direct either to Jermyn St or the rooms of the Geol Society.

The laird of Geanies (Sheriff Murray) desired me to express his sincere regrets at his being ignorant of your last visit to his district of Ross until it was too late to offer you the hospitality of his house; but hopes when you again visit your native country, you will make Geanies your headquarters;—and from my own experience, nowhere in the North would you be more gladly received & hospitably entertained.

P.S. I have had M[r] Lubbock[124] down among the [middens] of Moray, & he finds them strongly resembling those of Denmark. I am satisfied that some of them have hitherto been regarded as natural collection of shells as a mixed lunch.

To: Gordon
From: [Jermyn Street]
Dated: 19 April 1863
DDW 71/863/39

Yours of the 16 is very gratifying to me & I shall be delighted to lay the Sections (I hope with descriptions) of Mr Joass before the Geological Society.[125] The slab with footprints had better be sent there i e to Somerset House, though I should wish all the documents communicated to myself (Jermyn St).

I long to know if there be a conformity between the beds of Geanies & the strata of Tarbat Ness.

Or better I want to know if there be a mineral transition from the one set of beds into the other.

Between Port Mahomack & the Ness I can answer for it; but then the question comes—what is Port Mahomack?

Geanies with its Old Red fishes is a true base.

I have been somewhat shaken in my impression that the Upper Sands of the Murray & Dornoch firths are parts of the Old Red, by a few scratchings of Hugh Miller which Miss Allardyce[126] of Cromarty sent me & which are

---

[123]As mentioned in Part One, it is clear that Joass believed he was being of direct assistance to Murchison, and maybe he was. In his letter to Gordon dated 9 March 1863, he said: "I suppose I may now set to work at my sections for Sir Roderick for I think I can do nothing more to make my notes more complete."

[124]Sir John Lubbock, later Lord Avebury (1834–1913), banker, MP, naturalist. FRS 1858 (not to be confused with his father of the same name who was also a Fellow of the Royal Society of London).

[125]"On the Relations of the Ross-Shire Sandstones containing Reptilian Footprints" by the Rev. George Gordon and the Rev. J.M. Joass, *QJGS*, 19, (1863): 506–510.

[126]Resident of Cromarty and acquaintance of Hugh Miller. She was interested in the natural history of the north of Scotland. See footnote for DDW 71/863/50.

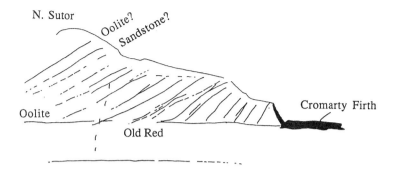

as seen on the flank of the N. Sutor.

But the point requires to be looked at & I much wish that Mr Joass or some one would do so before the field day comes off here.

If the facts be as above, mere apparent conformity proves nothing. We must have a true passage or transition. Otherwise we must strike and not claim so many <u>beasties</u> for our dear 'Old Red'.—

To: Murchison
From: Birnie
Dated: 21 April 1863
LGS/M/G10/4[127]

Along with this letter I have posted for you M^r. Joass' plan and section etc. of Tarbatness,—from <u>Geanies</u>, where the undoubted old red fossils are found, to the <u>Lighthouse</u>, where the reptilian footprints occur. I fondly hope that you will look on them as so interesting and so well executed, as to be laid by you before the Geol. Society—at an early meeting, and thus secure for M^r. Joass the credit of illustrating the connexion between the two points, before others come on the ground.[128]

Enclosed is Mr Js' description, which he has drawn up, and which, when taken along with the sections, will I trust induce most, if not all, geologists to acknowledge that, notwithstanding the included Reptilian footprints, you were perfectly correct when, so long ago as 1827, you were led 'step by step' through the strata without any evidence of a new formation coming on; and when, in 1858, you stated that the yellow sandstones of Morayshire are linked as to the red conglomerate and sandstones, containing *Holoptychius,* as they are to the cornstones of this tract, the whole forming a connected and natural series.

---

[127]The draft of this letter is DDW 71/863/46.

[128]The only significant difference between what was written in the draft and what appears here in the final copy is: "I fondly hope you will look on them as so interesting & and so well executed as will to be laid by you before the Geol Society—as our geol mentor and thus secure for us the credit of thus illustrating the connection between the two points before others come on the ground."

I shall be glad to hear how I am to dispose of the large flag, with foot-prints from Tarbatness, and alluded to by Mr. Joass in his paper, and referred to in my last note.

To: Gordon
From: [Jermyn Street]
Dated: 24 April 1863
DDW 71/863/42

Your dispatches including the map diagrams & Memoir of Mr Joass came regularly to hand & I will have great pleasure in laying them before the Geological Society on the 20th May.[129] For our next meeting on the 6th May other Memoirs were announced before your documents came to hand.

Your previous letters & other drawings etc. Mr Joass as sent to me last year will also be made use of.

I begged Mr Huxley to ask you to send the new slabs to this Museum.[130]

He Huxley abides firmly by his opinion of the Secondary age of his *Stagonolepis*.

After all the sectional & stratigraphical evidences which have been obtained from the Tarbet Ness promontory (which I was always convinced would not explain the Elgin & Burg Head strata) it is very difficult to dissociate the fish & reptilian beds. They evidently belong to the same <u>consecutive geological series</u>.

It may however be contended that the Reptilian beds are not of Devonian or true Old Red age, though Palaeozic.

Pray tell Mr Joass to whom I will write hereafter that I think he has done his task <u>admirably</u>.

To: Murchison
From: Birnie
Dated: 15 May 1863
DDW 71/863/50

Accompanied by Mr Stables Mr Joass and Major H. Drummond,[131] Bengal Engineers, I spent the whole of yesterday on and around the Nigg or North Sutor of Cromarty. We examined it minutely in reference to the sketch sent you by Miss C Allardyce.[132] This sketch represents the Oolites lying con-

---

[129]The report to the Geological Society was in fact made on 17 June 1863.

[130]Huxley had already done so, though not before 22 April which was the day on which Gordon sent the specimens to London by rail to Aberdeen for shipment by boat from there.

[131]Drummond had served in the 3rd Light Cavalry in India where he also published the results of his scientific work in the *Journal of the Asiatic Society of Bengal*.

[132]Resident of Cromarty and President of the Dorcas Society which distributed clothes to the poor. This is the Miss Catherine Allardyce to whom Hugh Miller refers in The Old Red

formably to the Old Red Strata, and as if they occurred on the <u>west</u>, or inner side of the North Sutor—thus

Unfortunately Miss Allardyce was at Inverness, but one of her sisters informed me that it was on the east or Moray Firth side of this Sutor that the Oolites referred to in her sketch that had been sent you were to be seen and a specimen of which was shown to me.

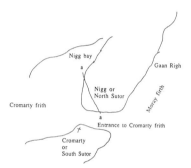

We examined all the bits of rock and strata that were to be seen cropping out on the line a a on the west side of the N. Sutor and found in them nothing but Old Red Sandstone. We then proceeded by boat down the Moray firth as far as Shandwick landing at several places & walking along almost the whole shore spending a considerable time at Gaan Righ.

Here the Oolite beds first appear, covered however at high tide, lying at the base of a cliff of about 250 feet high and are clearly an extension of the larger and well known deposit of Shandwick. The tides were low & very favorable for our work. The complete unconformity of the Shandwick Oolites to the Old Red Sandstone is acknowledged on all hands. The seeming conformity of the Oolite beds at Gaan Righ to the O.R. arises from the accidental change of dip (caused by a fault or rather slip or slide) in a larger mass of the rocks (Old Red) in the overhanging cliff. The strata of this mass if projected on their present plain would most probably <u>overlie</u> the Oolites. The persistent northern dip of the Old Red strata in this line of coast all the way down to Shandwick comes out distinctly when seen from the boat

---

Sandstone as "a lady who, to a minute knowledge of not a few other branches of natural science, adds an intimate acquaintance with the fossils of our northern formations" (*The Old Red Sandstone*, 12th Edition, 1869, Edinburgh, p. 361–362).

when sailing in the offing. Faults or slips there are in some places but they do not present any difficulty. The point called the Actnach head may be thus (rudely) represented.

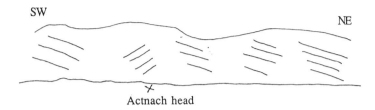

Actnach head

The Actnach head (or Red Nose) being a tumble down or slip rather than a fault strictly speaking.

I am therefore convinced that the great question at issue the age of our reptiliferous beds is not in the least affected by any appearance presented by the Oolites at Gaan Righ—the spot I believe alluded to by Miss Allardyce in her note to you.

P.S. Since writing the above Mr Joass' sections have just come to me by post and his final will illustrate the locality better than any here. G.G.[133]

To: Gordon
From: [Jermyn St]
Dated: 20 May 1863
DDW 71/863/54

Your [   ] & sketches have been duly received. It had been arranged that they & all relating to Tarbat Ness & Cromarty would come off tonight.

I found however that they had put 3 papers before <u>our</u> Communication (for I have herded the Memoirs & letters into a preface)[134] and hence I have requested that a point of such very great importance should not be <u>booked</u> at the tail end of an Evening.

It will therefore be No 1 & receive every attention on <u>Wednesday June 3rd.</u>[135]

By referring to my old notebooks & descriptions I find that there never was anything so clear—(indeed it is as printed in the GeologL Transactions) as the <u>total unconformity</u> of the <u>Liass</u> & Oolite? of the coasts of

---

[133]This sentence was added later by George Gordon: "(To Sir R. Murchison 15 May 1863 about Oolites at Gaan Righ near Shandwick)."

[134]The joint article by Gordon and Joass was preceded by an introduction by Murchison in which he set their contribution into the context of his own work on the geology of Ross-shire spanning more than 35 years (see Gordon and Joass, 1863, On the relations of the Ross-shire sandstones containing reptilian footprints, *QJGS*, 19, 506–509).

[135]The paper was presented on 17 June 1863, as already indicated.

Cromarty & Ross to the Old Red Sandstone. Hugh Miller though a good naturalist & a splendid writer, was not a stratigraphical observer & the less that is said about it for the sake of that remarkable man the better.

I am so busy with the discovery of the Nile source (see Times of Monday last) & in preparing for my great Geographical Anniversary of Monday next[136] the 25 that I am glad <u>our</u> papers are deferred—Mr. Joass' sketches & sections are admirable.

To: Gordon
From: 16 Belgrave Square
Dated: 18 June 1863
DDW 71/863/62

Though I am specially overwhelmed with business, I seize the first opportunity to tell you that the Tarbat Ness Memoir was read first at our last meeting of the Geologists & had <u>full</u> & <u>ample fair play</u> and <u>was much approved</u>.

Ramsay in the chair called on me to give first a general view of the whole subject & then to dwell on the doings of Yourself & friends in bringing out the clear evidence in question.

(Somehow or other the graph paper with your blue & red lines was the only thing wanting).

They tell me that I never was more successful in making it clear through the witness of Mr Joass Mr Campbell & yourself that there is a perfect mineral & geological, conformable transition from the true Old Red up into the Reptilian Sandstones, & I left Huxley to say whether it was not possible, that the footprints, of which drawings were exhibited might not have been made by an animal of the Carboniferous Period?—for all that I certed, is, that all the upper beds are <u>Palaeozoic</u>. He made one admission which you will be delighted to hear, viz, that the big foot print of Tarbat Ness is identical with one from Cummingston & the Elgin shore. So now we have the proof of what myself & Ramsay contended for as to the unbroken sequence in your Country.

Ramsay summed up well in favor of true stratigraphical proof & Huxley faltered in "standing to his guns".

He simply relies on the *Stagonolepis* & his <u>prints</u> as having Triassic character.

The Triassic folks will have an additional difficulty now that we know that all the Lias & Oolites are utterly unconformable to the Paleozoic rocks in question.

Again, thanking yourself, Mr. Joass, Mr Campbell, your friend the military man whose eye seems to be good.

---

[136]As part of his duties as the President of the Royal Geographical Society, Murchison presented an annual address summarizing the advances in the field during the previous year.

To: Gordon
From: [Jermyn Street]
Dated: 13 July 1863
DDW 9/1[137]

If I did not reply to your letter of the 4th July with its enclosure of a list of the animals which live on the shores of the Murray Firth, I regret it, provided the loss of a week should have been of importance.

There never was a clearer case in Natural History than that which you have set before me.

All the Echinoderms, Crustaceans & Mollusks which you have detected & named, are unquestionably & <u>altogether</u> marine.

So says Professor Huxley & so indeed must every Zoologist say, including Professor Owen.

I venture to hope that the valuable data you have collected will enable our legislators to exclude the Moray Firth from riverine & fresh water conditions.

To: Murchison
From: Birnie
Dated: 16 July 1863
Interleaved in the volume entitled "Scientific Papers by Reverend Doctor George Gordon" and catalogued as DDW 9/1

I beg very gratefully to acknowledge the receipt of your kind note of 13, the contents of which will give me much confidence when expressing the opinion I had formed of the waters of the Inverness Firth.[138] Professor Harkness was here for two days last week & also yesterday (after his return from Ross-shire). He [   ] the pains you have taken on both sides of the firth—with the addition of adopting a view which for some time back has been pressing itself upon me, namely that in the Elginshire sandstones & Silurians we have a series of faults (from SE to NW) which bring the limestones always to the top: thus Glassgreen, Linksfield, Spynie, Lossiemouth, Inverugie, & Cothall are all on one and the same plane of deposition.[139]

I was first lead to this view by not finding the limestones between Alves & Blenie where they must have appeared, I think, had they been underlying the Quarrywood range. It will not, indeed, be conclusive that we have

---

[137]This letter has been included for the sake of completeness of the Gordon-Murchison correspondence even though it has no bearing on the geology of the region.

[138]Gordon here refers to a legal problem on which his advice had been sought; that is, the identification of the point in the Moray Firth at which the laws governing fresh water gave to those governing the sea.

[139]The geologic map of the Elgin region show a series of normal faults which trend NE-SW with the downthrown block on the southeast side. Thus, the stratigraphy is repeated as can be observed in the Quarrywood ridge and along the coast. The cornstone bed which is a limestone is a widespread unit that can be used as a marker bed.

still but one other cornstone above that of [   ] untill we shall have produced reptile remains from below the limestones of Cothall, Glassgreen
etc., as well as at Spynie and Lossiemouth.

To: Murchison
From: Birnie
Dated: 29 December 1865
LGS/M/G10/5

Permit me to offer you my warmest congratulations on your receiving the
merited honour[140] which our gracious Sovereign has just been pleased to
confer on you,—and to express my heartiest wishes that you will have
many happy returns of the Season to enjoy it.

Nothing new or interesting has of late turned up in the geology of my
native Province,—otherwise the Authorities at Jermyn Street should have
heard of it.

To: Murchison
From: Birnie
Dated: 24 July 1866
DDW 71/866/26[141]

Having promised to report to you any discoveries that might be made in
our upper Sandstones, it gives me pleasure to inform you that another
specimen of the *Telerpeton Elginense* (so far as I can read it) has been found
at Lossiemouth, and is now in the possession of Mr Grant[142] Asst Teacher
there Mr G. who is happily in the spot & has an eye & zeal for the work.
Last week as soon as he heard that the quarrymen had seen something
new, immediately rescued the specimen from further injury for it had already been broken & has preserved for the use of science—Some parts of
the specimen particularly what I take to be the cast of the interior of the
skull are more distinctly brought out than in *Telerpeton Elginense.*

Another, and but the second specimen of this ancient reptile has been
lately extracted from our upper Sandstones. It is now some fifteen years
since the first specimen was found in the quarry at Spynie & brought to the
late Patrick Duff Esq. of whose cabinet it long remained the chiefest gem.
It is now in the possession of James Powrie Esq of Reswallie Forfarshire.[143]
Some of our readers will perhaps recollect what a [sustained] and [severe]

---

[140]According to Geikie, Murchison was elevated to a baronetcy in 1866 (Geikie, II, p. 330).
[141]This draft has the notation 'Telerpeton Elginense' at the top.
[142]Mr. Grant was a teacher at the General Alexander School in Lossiemouth.
[143]James Powrie of Reswallie (1814–95) J.P., F.G.S., F.R.S.E. An energetic geologist and collector of specimens.

strife were caused by the introduction of this fossil to the Scientific World. Questions arose some of them [are] not settled. It was figured in the Illustrated News & is now a well known figure in every illustrated Geol. Text Book. It is no longer unique. Last week in building a house at Branderburgh some of the workmen had got a block of sandstone from the Lossiemouth quarry, & which on being broken up betrayed some traces of a small skeleton, or rather cast of a skeleton. It was presented to Mr Grant of the Genl. A. Schools & is now in his keeping where it will be properly preserved & exhibited for the use of Science for no doubt many of the paleontologists of the South will be anxious to examine & illustrate it by pen & pencil.—This discovery does not settle the geological age of the Lossiemouth beds as a scale of a true Old Red Sandstone fish if found in them would—it however adds to the evidence showing that in their formation the hills of Lossiemouth & Spynie repeat each other—in the specimen from Spynie being in the possession of the late Mr Duff & now in the cabinet of Mr. Powrie. This second sp. is from the upper siliceous beds at Lossiemouth & not in those underlying softer strata where the *Stagonolepis* have hitherto been found. This discovery strengthens the evidence already obtained that the Spynie beds are but a repetition of those at Lossiemouth.

Mr. Grant has also a specimen shewing the teeth of the *Hyperodapedon* in their fragmented form.

I trust others will be found in the course of the next twelve months when the whole will be exhibited at Dundee.[144]

To: Gordon
From: Erlwood, Bagshot
Dated: 30 July 1866
DDW 71/866/25

Your letter of the 24th was very welcome; & of course I rejoice in the discovery of another *Telerpeton*.

But Huxley is vexed to hear, that it is to remain unexamined <u>for more than a year</u>.[145]

Is it not worthwhile to have it described, *ad interim*, & returned to the proprietor when he or you can exhibit it at Dundee?

The Duke of Buccleuch has consented to be named President for Dundee meeting.

Hoping that we two old stout fellows may meet there.

P.S. Lady Murchison being in very poor health, I have taken this house for her during the summer & do not intend to go far from her—only to Nottingham & a visit or two in England.

---

[144]The British Association met in Dundee in 1867.

[145]But it did not remain unexamined for more than a year as Huxley read "On a New Specimen of *Telerpeton Elginese*" to the Geological Society on 19 December 1866.

To: Gordon
From: Erlwood, Bagshot
Dated: 13 September 1866
DDW 71/866/37

I was of course rejoiced to hear of an addition to the reptilian Fauna of your Elgin Sandstones, but I am disappointed, or rather I shall be so, if the discoverer will not allow us to examine & describe the creature before the Dundee Meeting in 1867.

I say this with some feeling, inasmuch as [I] am preparing a new Edition of *Siluria* in which I wish naturally to render the Elgin chapter more perfect in which the long appendices B & Q will be worked into the main text of the volume together with all new acta.

Among the latter I hope to get something from Mr Joass.

Pray see if you could induce the Scientific Schoolmaster to let his new animal be sent up to Jermyn St.

I shall live here till the middle of October, having taken the house for my wife's health, but I go to & fro London when occasion requires.

To: Gordon
From: Murchison
Dated: 24 January 1867
DDW 71/867/3

A great change has come over me! and '*mirabile dictu*' I have abandoned altogether the cause of the Reptiliferous sandstones of Elgin & Tarbet Ness being part of the <u>Old Red Sandstone!!!</u>

A discovery made by Huxley from an identification of one of your heads and teeth of the *Hyperodapedon* with a fossil from Warwick has settled the business.

It is a curious fact that this *Hyperodapedon* of Warwick comes from the beds of the Upper New Red Sandstone, which in a joint paper with the late Hugh Strickland[146], I shewed (30 years ago) to be the equivalent of the German 'Keuper' and above the "Bunter Sandstein". We also gave engravings of the foot tracks of a reptile near Warwick, but never found one of the animals. This fossil sent to Huxley by a Mr. Lloyd of Warwick, and the <u>additional fact</u>, that the bed which connects the Keuper with the Lias (or rather separates the two thus is naturally connected with the former,

---

[146]"On the Upper Formations of the New Red System in Gloucestershire, Worcestershire, and Warwickshire, showing that the red (saliferous) marls, with an included band of sandstone, represent the Keuper or "marnes irisées," and that the underlying sandstone of Ombersley, Broomsgrove, and Warwick is part of the "Bunter Sandstein," or "grès bigarré," of foreign geologists", by R.I. Murchison and H.E. Strickland, 1837, *Transactions of the Geological Society of London*, second series, v.5, p. 331, and *Geological Society Proceedings*, v. 2, p. 563.

the now so called Rhaetic) exists at Linksfield has completed my conviction. I now see, that the enormous quantities of Lias fossils with which your region abounds, had in these Uppermost Sandstones of the Trias or New Red a <u>natural foundation</u>.

I have therefore erased the *Telerpeton* from my woodcuts for in *Siluria* I have really to do with <u>Mesozoic deposits</u>.

I explain in my book that up to the very last moment I was going to contend for the Old Red antiquity of your Reptilian beds and had actually penned two or three pages of reasoning in favor of that view, mainly based on the undeniable fact of the conformity assumed by so many good geologists including yourself, Harkness Joass etc. But in truth, all must give way to such fossil evidence. So long as all the Reptiles of your tract were <u>unique</u>, I stood up for the old argument, but when your *Hyperodapedon* is found at Warwick & that Rhaetic beds with fossils occur at Linksfield, I am obliged to strike and connect the hitherto anomalous strata with the overlying Lias & Oolites of the Murray Firth & Brora described by me 40 years back.

It is moreover curious that in my first paper (written in <u>1826</u>) I did speculate on <u>New as well as Old Red extending into the fertile plains of Easter Ross</u>, my own hinter land—& that I then said 'that in a region where there was no dividing Carboniferous series, the one formation might seem to follow the other; i.e. the <u>new</u> to follow the <u>old</u> & often exhibiting great similarity of lithological character, and be [difficultly] distinguished. On the whole I rejoice at the discovery. It suits my creed as to the succession of geological creation.

To: Murchison
From: Birnie
Dated: January 1867[147]
DDW 71/866/8

As the Bedford motto runs, & in reply to the letter with which you have favoured me, I say *'Che sara sara'*. —The exciting interest, so long & pleasantly kept up—as to the geological age of our upper, or rather reptiliferous sandstones, if not extinguished, has now dwindled down to that which might be stirred by such subordinate questions as these;—1. At what point in the well exposed section from Geanies to Tarbat Ness, does the break occur, where the Triassic beds (Keuper) come on upon the Old Red?—2[148] How comes such a break to have been overlooked, or misinterpreted, by so many? 3. If this break be not marked or discoverable, is

---

[147]This letter is a reply to Murchison's letter of 24 January 1867 and therefore cannot have been written in 1866 despite the catalog number. Murchison responded to this letter on 31 January 1867.

[148]The rough draft (DDW 71/867/34) to this letter has the number two at this position.

there another instance of such a lengthened epoch of quiescent hori-
zontality,—as must have prevailed in the region of the Moray firth—
while the <u>Carboniferous</u>, the <u>Permian</u>, & the Lower <u>Trias</u>, were being deposited? i.e. from the era of the Bishopmill beds to that of the Findrassie,
or, from that of the Geanies beds, to those of Tarbat?—4. Is it not remarkable, that, so far as I have observed, and I think you will concur with
me, wherever the undoubted <u>Oolites</u> & <u>Lias</u> are seen <u>in situ</u>, as at Shandwick, there is evident and great <u>un</u>conformity to the underlying strata.
This is the reverse of what might have been looked for,—there is unconformity between the Lias & the Trias, systems otherwise far more nearly
allied to one another than the <u>Trias</u> & <u>Old Red</u>, between which last however there is in the Moray firth so well marked a conformity. or 5, Does
this structure of the Moray firth give any countenance to the idea that I
believe was once entertained by some, that the <u>Coal</u> formation was an
episode—or byplay, formed, in inland seas like the Caspian, at the very
time that the Permian etc Strata were deposited in the ocean shores? 6.
Perhaps a more close examination of the Pentland firth, the uppermost
sandstones of which you classed in 1826 along with Tarbatness as
"Newer Red Sandstone", would find out the point where the break, if
break there be, occurs between the Old Red & New—the Devonian and
the Triass (Keuper).[149]

The fact that Profr Huxley has found the *Telerpeton* to be nearest, in its
organisation, to the lizards that now run about, than to any Paleozoic or
even Mesozoic remains does not *ex necessitate*, upset the old argument
from stratification. It may however be otherwise, when we consider what
he has discovered as to the *Hyperodapedon* from Warwick.

Although the geologic age of our reptiles will now, I suspect, be no
longer considered an open question, I feel that a mesozoic mollusk or fish,
from the Lossiemouth bed, would be to me a still more convincing proof.
Would it then be proper to ask a grant from the British Association when
it meets at Dundee, for the purpose of quarrying the reptile stratum
there?[150] This stratum is so friable that the workmen, finding it worthless
for building, have left much of it in an accessible state. I should not un-
dertake alone the responsibility of expending the Associations' money; but
would be glad to be a member of a local Committee. Meantime Mr Grant

[149]The Pentland Firth separates the Orkneys from the mainland of Scotland. Murchison first
visited the region in 1826 and returned with Sedgwick in 1827. In his paper resulting from his
1826 trip he speculated that the Pentland sandstones were the same age as the red conglomerates of Sutherland and Ross-shire. In their paper resulting from the 1827 trip, Sedgwick and
Murchison speculated that the sandstones of Dunnet Head and Tarbat Ness belonged to the
New Red Sandstone. They recognized that the Caithness Flags overlay the Old Red Sandstone proper and speculated that the fish-bearing units were equivalent in age to the coal-
bearing Carboniferous further south.
[150]Gordon received a grant from the Government Grant Fund of the Royal Society in 1867
(see DDW 71/867/25). There is no record of a grant from the British Association having been
made to support further excavations in search of fossil reptiles in the Elgin sandstones.

will be on the outlook for whatever the workmen may chance to turn out of this now rather famous locality.

To: Gordon[151]
From: [Jermyn St]
Dated: 31 January 1867
DDW 71/867/8

Nothing more apposite than that when we meet at Dundee, you should or if you like I will ask that a grant be assigned to you & Mr Grant, to work out the Lossiemouth fossils.

As to your doubts & queries, I think that most of them are not difficult to explain.

1. Why we did so ill before?

—Answer. In early days we worked much more loosely & had no <u>fossil guides</u> to make us suspect or look out—Next, Ramsay & myself did not test the critical part of the Ross-Shire Section.

2. As to the <u>length of time</u> in which an old formation may remain <u>perfectly quiescent</u> until a much younger one is deposited on it, <u>I have no difficulty whatever</u>. In the N East of Russia in [   ], the Carboniferous limestone is perfectly flat & in several places is at once covered <u>by equally horizontal</u> deposits with Arctic Shells some of them of existing species! This is a greater gap with a vengeance!

3. I agree with you that the difficult point is the evident discordance between the Oolites & the Old Red of Cromarty. But here again, this only proves that after the deposition of the <u>Bunter & probably the Lias</u> upon the Old Red, a great commotion shook the Oolites athwart some of the older rocks. In short the disturbance at the close of the Oolite Series as found at the north end of the Brora district was tremendous, for we have there breccias made up of every bed in the Series.

I believe that the Coal formation never had any existence, or any representative in the north & lastly I am now certain that the true Upper Old Red Sandstones of Dunnet & Hoy Head & the Orkneys are <u>not</u> equivalent of your Reptiliferous Sandstones for they contain true Devonian land plants etc.

This discovery also opens out to my vista another satisfactory explanation of certain soft, marly & in every respect Triassic-looking Red Sandstones of the West Coast, which Sedgwick & Self (1827) suggested to be New Red & which though severed from, are still in, the region of grand Liassic shell deposits.

---

[151]This letter is the original from which DDW 71/867/6 was copied. There is evidence to suggest that Gordon made several copies of this letter.

I had I assure you prepared all the possible arguments for the Old Red view, including the very strong facts of the recent occasion of land plants & trees for our lizards as brought out by Dr. Dawson of Montreal.[152]

But I have succumbed—the more so as the Rhaetic fossils of the intermediate beds between the Keuper Sandstone & the Lias have been found at Linksfield.

Coupling this with the abundance of Lias remains to the east of Elgin I see no possibility of evading the supraimposition of the Trias to the yellow Old Red of Elgin. N.B. Much in my opinion may be done during the lowest tides in examining those scars & islets of Sandstone etc which fringe the coast of Lossiemouth & Burgh Head.

Ramsay & Self found some casts of shells in these, & if any Lias types could be got there it would be valuable.

As far as I could judge the strata were very broken & dipping variously.

To: Gordon
From: James Nicol
Dated: 31 January 1867
DDW 71/867/7

I am much indebted to you for your most interesting communication. Sir R's letter is highly characteristic. It would be hard to find a discovery that did not suit his views or a view that had not been indicated in some form or other in his papers on this region. Ever since the *Stagonolepis* was proved to be a reptile he has been vacillating between the fossils and the strata. Could you get him to Elgin for a couple of days to look at the beds—they would be Old Red, at least till he saw Huxley.

We must however hear more of the exact nature of Huxley's discovery before judging of its value. Is the identity generic only or also specific? Even if the latter, there still remains the deeper question. Is one identical fossil sufficient evidence to overrule and over ride all evidence from stratification.

The points you note are very good. But is there any point where lias rests on undoubted Reptilian beds? Lossiemouth? Shandwick is high in the Oolites and what are the sandstones below? Had we undoubted lias unconformable on true Reptilian your point would be very strong. But as you say '*Che sara sara!*' A few shells or fishes may any day alter all our notions—even Sir R's new views. Meantime having no special theory of the order of Creation to defend, I can wait with all patience for more light and any facts that may come.

I this forenoon sent on yours with enclosures for Mr. Powrie.

Mrs Nicol joins me in kindest regards to Mrs Gordon, the young ladies and yourself. We are glad to see you have come safe through this terrible storm.

---

[152]John William Dawson (1820–1899) was a Canadian geologist and principal of McGill University from 1855 to 1893. He was the first president of the Royal Society of Canada and was knighted in 1884.

To: Gordon
From: [Jermyn St]
Dated: 7 November 1867
DDW 71/867/25

Our friend Professor Huxley having reminded me that the Royal Society had a fund at its disposal,—the Government Grant Fund—& that is would be very desirable to obtain from it £50 to be placed at your disposal to dig up <u>more</u> Reptiles, or other animals in the Elgin Sandstone, I wrote to apply for the same & Dr. Sharpey the Secy.[153] informs me that it was granted & you will receive from the Asst Secy Mr White a cheque for the same.

I hope you got your copy of the 4 Edn of *Siluria* which was sent to you by the Publisher at my request.

I now enclose a fly page to be pasted in on the inside of the binding as a prelude.[154] Allow me to suggest that some endeavour should be made to find the remains of the animal that formerly walked about so frequently in the Yellow Sandstones of Easter Ross.

---

[153]William Sharpey (1802–80), Professor of Anatomy and Physiology at University College, London. FRS 1839. Secretary to the Royal Society 1853-72. Murchison's letter of application to Sharpey to enable Gordon to "further open up the Old Red" is in the Library of the Royal Society of London (MC.8.99). Though the catalogue date is 20 October 1869, the 9 should almost certainly be a 7.

[154]The notice referred to was written 30 October 1867 and is as follows: The reader of this edition will find that a very important change has been made in my views as given in former editions, respecting the age of the Upper Sandstones of Elgin and Ross-shire, which I have hitherto classed with the Devonian or Old Red Sandstone. My previous conclusion was founded entirely on the strong natural evidence presented, to me, by the conformable superposition of those beds to the strata of the inferior and unequivocal Old Red Sandstone replete with its well-known fossils. This opinion was confirmed by the examination of the rocks in question by Professor Ramsay, Professor Harkness, the Rev. George Gordon, the Rev. J.M. Joass and others.

The existence, in strata of Devonian age, of reptiles of so high a class as the *Telerpeton* (see fig. 73 in my last edition, p. 289) and the *Stagonolepis* was not, indeed admitted by me without great reluctance, inasmuch as, if eventually substantiated, it would have weakened the main argument that runs through all my writings, which shows a regular progression from lower to higher grades of animals, in ascending from the older to the younger formations. Most joyfully, therefore, did I welcome the remarkable identification by Professor Huxley of the *Hyperodapedon* of the New Red Sandstone of Warwickshire with the *Hyperodapedon* of Elgin; and bowing, as I have always done, to clear paleontological proof, I have now excluded all that portion of my former editions which placed these reptiles in the Old Red Sandstone.

The importance of this rectification, due to my eminent associate, has very recently received a wide extension; for among the fossil remains collected in India by the late Rev. S. Hislop, Professor Huxley has also found the *Hyperodapedon*.

The formation in India containing this reptile has been considered by Professor Oldham, the Director of the Indian Geological Survey, to be either the Trias (New Red Sandstone) or the representative of an intermede between the Palaeozoic and Mesozoic rocks. In all probability this correlation will have to be extended to South Africa, since one of the characteristic fossil reptiles of that country, the *Dicynodon*, has been found in the Ranigunj beds of this age in India.

I take this opportunity of further stating that I have not adverted in the Preface to a great number of important additions which I have made in this edition; they are, in fact, so numerous that, if a smaller type had not been used, the work would have been swollen to an unreadable size.

To: Murchison
From: Birnie
Dated: 8 November 1867[155]
DDW 71/869/25

I feel much gratified by the grant of £50 given at your insistence by the Royal Society. I shall endeavour to lay it out to the best purposes.

Quarrying at random would be expensive & I fear disappointing. The best plan will be to instruct the workmen, and when they come upon any fossil, to pay them well for extracting it carefully.*[156]

I have not yet had the pleasure of seeing the 4 Ed^n of Siluria but I have no doubt the acceptable gift will come by and bye. Many thanks for the flyleaf.

To: Murchison
From: Birnie
Dated: 27 November 1867
LGS/M/G10/6

I have just had the honor of receiving your valued gift,—a presentation copy of the 4th Edition of 'Siluria'.

Be assured it will be cherished as an heirloom, highly esteemed intrinsically & extrinsically.

To: Gordon
From: James Nicol
Dated: 2 December 1867
DDW 71/867/30

I see in Saturday's Scotsman a notice of a new Reptile in the Elgin beds. It is said to decide the fact of great crocodile-like beasts having lived in the Old Red Period.

I am not you know at all averse to this notion and should be glad to have some certain proof of the fact. But there are one or two points on which I should be very glad to have a little more information if you would kindly give it me.

1st Have any old Red fossils been found in the same beds or in beds certainly higher. Unless this is the case, their old Red age will still be disputed.

---

[155]Certainly 1867 not 1869 because the answer to 867/25. Gordon received the fourth edition of Siluria long before 1869.

[156]The asterisk is Gordon's and refers to an almost indecipherable addition at the foot of the page, which must have been designed to reassure Murchison that "the operation of the quarry will last for a year or two at least."

Even if fish found in lower beds in the same quarry we shall have no end of doubts.

2nd Are the remains truly Reptilian. In the notice the teeth are spoken of as in "double rows" which is not the case in the Crocodiles and is rather a fish character. I am afraid from this that the jaw may turn out to be that of a *Holoptychian* fish and not a true Crocodile. These (*Rhizodus*) have large tearing teeth mixed with smaller ones. Then they are Old Red and thus quite in their true place.

Whereabouts is the Quarry? There is one of white stone near the station. And I think you took me to some others, but I have not had time to turn up my notes.

When we left Strathspey last autumn we went to Keith, intending to stop for a time. But the inn was so dusty and uncomfortable looking that we were glad to get out of it—and drove over to Cullen. There I found a fine series of the quartz rocks. I have no doubt about their true position in the series. But it is of no use speaking to those who will not hear.

To: Gordon
From: James Nicol
Dated: 5 December 1867
DDW 17/867/30

I got your note and also this morning a copy of the Elgin *Courant* for both of which accept my best thanks. It was the mixture of fish and reptile characters in the Inverness notice that puzzled me. I have no fear of a Crocodile even in the Old Red—but I did not like his double row of teeth.

Your note however puts it all right—but leaves the problem where it was. There will be denial so long as denial is possible. Murchison has no opinion of his own—oscillates just as those he last consults. Had we him in Elgin for a few days he would be round to Old Red—even with ??

I got a copy of *Siluria* sent me, but have not had time to look into it as yet more than in a very cursory manner.

I am not quite ignorant of the Mulben section. I walked along the ravine—by the Burn and railway and also by the road when at Rothes some years ago. But I should still like to have your guidance through it. What I specially want is a good junction of the quartzite and gneiss. I have my own notions. But sections beyond dispute are not so readily got. You have so much drift in all that region of Moray.

I should have been glad of your company from the Spey to Glen Roy. I had a fine drive up to the foot of Corryarick whence Mrs N. drove back to Kingussie. I then walked across by Glen Roy & [Glowy] to Spean Bridge, where they had no quarters for a poor pedestrian. So I had to tramp on four miles in the dark to Roy Bridge. Where being better known I got a hearty welcome. If you have not seen Glen Roy you have still a treat. Like your Culbin sands it is quite unique. I am still all for Lakes. But unless I were in London to take my own point—there is no use wasting pen and paper on the matter.

Like you we have had a most severe storm—but no fine flowers to waste. It threatens to be a very hard winter, and trade here very depressed. Then potatoes have quite failed.—scarce one that is eatable to be got.

APPENDIX I
SELECT LIST OF MURCHISON'S ARTICLES
RELATING TO THE NORTH–EAST OF SCOTLAND

1 1827. On the coal–field of Brora Sutherlandshire, and some other strat-
ified deposits in the north of Scotland, *Trans. Geol. Soc. Lond.,* Series 2,
2, 292–326
Plate XXXI – Fig. 1: map; Fig. 2: transverse sections of the east coast of
Sutherland, Cromarty and Ross; Fig. 3: sectional view of Ross and
Cromarty; Fig. 4: coast of Skye.
Plate XXXII – sketch of rock immediately above the Brora coal.

2 1827. On the Old Red conglomerates and other Secondary deposits on
the north coast of Scotland, *PGSL,* 1, 77, (with Adam Sedgwick).

3 1828. On the structure and relations of the deposits contained between
the Primary rocks and the Oolitic Series in the north of Scotland, *Trans.
Geol. Soc. Lond.,* Series 2, 2, 125–160, (with Adam Sedgwick).
Plate XIII – Geological sketch map of northern Scotland.
Plate XIV – Fig. 1: Section of north coast of Caithness; Fig. 2: Section
from the Ord of Caithness to Dunnet Head; Fig. 3: Section from Ben
Wyvis to the North Sutor of Cromarty; Fig. 4: Section from Speyside to
Tarbat Ness to Sutherland; Fig. 5: Section through the Maiden Paps and
Scarabins.
Plate XV – Figs. 1, 2, 3: *Dipterus macropygopterus;* Fig. 4: Genus *Dipterus.*
Plate XVI – Fig. 1: *Dipterus Valenciennesii;* Fig. 2: *Dipterus macrolepidotus;*
Fig. 3: *Dipterus Valenciennesii;* Figs. 4 & 5: *Dipterus macrolepidotus;* Fig. 6:
Trionyx; Fig. 7: *Dipterus.*
Plate XVII – Figs. 1, 2, 3: *Dipterus brachypygopterus.*

4 1828. Supplementary remarks on the strata of the Oolitic Series, and the
rocks associated with them, in the Counties of Sutherland and Ross,
and in the Hebrides, *Trans. Geol. Soc. Lond.,* Series 2, 2, Part 3, 353–368.
Plate XXXV – Fig. 1: Map of western Scotland and the Hebrides showing the distribution of the oolitic series; Fig. 2: Section of Skye and adjacent islands; Fig. 3: Section of the south coast of Mull.

5 1829. On the geological relations of the Secondary strata in the Isle of
Arran, *Trans. Geol. Soc. Lond.,* Series 2, 3, 21–36, (with Adam Sedgwick).
Plate III – Sectional view of the Isle of Arran.

6 1839. *The Silurian System,* (3 vols.) London: John Murray

7 1841. Fishes of the Old Red Sandstone, *BA,* 1840, Glasgow, 99

8 1842. Anniversary address of the President, *PGSL*, 3, 637–687

9 1851. On the Silurian rocks of the south of Scotland, *QJGS*, 7, 137–68.
   Pp. 168–69: abstract of Part II of this article.
   Pp. 170–78: list and description of Silurian fossils compiled by J. W.
   Salter.
   Unpaginated: 3 plates of Silurian fossils.

10 1854. *Siluria. The History of the Oldest Known Rocks Containing Organic
    Remains, with a Brief Sketch of the Distribution of Gold Over the Earth*, Lon-
    don: John Murray

11 1855. On the relations of the crystalline rocks of the north Highlands to
    the Old Red Sandstone of that region, and on the recent discoveries of
    fossils in the former by Mr. Charles Peach, *BA*, 85.

12 1857. The quartz rocks, crystalline limestones, and micaceous schists of
    the northwestern Highlands of Scotland, proved to be of Lower Sil-
    urian age, through the recent fossil discoveries of Mr. C. Peach, *BA*, 82

13 1857. On the crystalline rocks of the north Highlands of Scotland,
    American Association Proceedings, Part 2, 57

14 1858. On the succession of the older rocks in the northernmost counties
    of Scotland; with some observations on the Orkney and Shetland Is-
    lands. Abstract entitled: "On the succession of rocks in the northern
    Highlands, from the oldest gneiss, through strata of Cambrian and
    Lower Silurian age, to the Old Red Sandstone, inclusive", as read on 3
    February 1858, *QJGS*, 14, 501–504.
    With the abstract, a note: "The publication of this paper is postponed".

15 1858. Some results of recent researches among the older rocks of the
    Highlands of Scotland", *BA*, 94

16 1858. On the succession of the older rocks in the northernmost counties
    of Scotland; with some observations on the Orkney and Shetland Is-
    lands, *QJGS*, 15, 353–417
    Pp. 418: Explanation of Plate XIII.
    Unpaginated: Plate XIII, facing page 419.
    Pp. 419*–421*: Explanation of the Geological Map of North of Scotland,
    followed by Plate XII: "First Sketch of a New Geological Map of the
    North of Scotland", dated 1859, facing page 359. Asterisks may indicate
    the later addition of this map and note and accounts for overlap of pag-
    ination with next paper.
    On p. 359 is a note: "This map will be issued in a subsequent Number
    of the Journal, it being my intention to revisit the Highlands this sum-
    mer, in order to satisfy myself on some still doubtful points, — partic-
    ularly in respect to the Reptiliferous sandstones of Moray".

17 1858. On the sandstones of Morayshire (Elgin, etc.) containing reptilian
    remains; and on their relations to the Old Red Sandstone of that coun-
    try", *QJGS*, 15, 419–436.

Pp. 437–39: Table 1 "giving a synoptical view of the Old Red Sandstone of Britain and the Devonian rocks of Devonshire and the Continent, with their characteristic fossils, and the list of the fossil fishes of the Old Red Ichthyolites of England, Scotland and Russia".
Pp. 435–37: "Postscript" dated 12 June 1859.

18 1859. *Siluria*, Third Edition, London: John Murray

19 1859. Supplemental observations on the order of the ancient stratified rocks of the north of Scotland, and their associated eruptive rocks, *QJGS*, 16, 215–240

20 1859. On the geological structure and order of the older rocks in the northernmost counties of Scotland, Murchison's address at the 1859 meeting of the British Association in Aberdeen, as reported in the *Elgin Courant*, 23 September 1859. See Appendix 6.

21 1861. On the altered rocks of the western islands of Scotland, and the north–western and central Highlands, *QJGS*, 17, 171–232. Pages 228–232 are an appendix countering a recent publication by James Nicol.

22 1861. On the coincidence between stratification and foliation in the crystalline rocks of the Scottish Highlands, *QJGS*, 17, 232–40

23 1861. *First Sketch of a New Geological Map of Scotland, With Explanatory Notes*, Edinburgh: W. & A. K. Johnston, and W. Blackwood & Sons, 22 p. with map (with A. Geikie).

24 1863. On the relations of the Ross–Shire sandstones containing reptilian footprints, *QJGS*, 19, 506–509, Murchison's introduction to the paper by Gordon and Joass (see number 11: Appendix II).

25 1864. On the Permian rocks of the north–west of England, and their extension into Scotland, *QJGS*, 20, 144–165

26 1867. *Siluria. A History of the Oldest Rocks in the British Isles and Other Countries; With Sketches of the Origin and Distribution of Native Gold, the General Succession of Geological Formations, and Changes of the Earth's Surface*, 4th Edition, London: John Murray

27 1869. Introduction to the Rev. J. M. Joass' notes on the Sutherland gold–field, *QJGS*, 25, 314

28 1869. Observations on the structure of the north–west Highlands, *Edinburgh Geological Society Transactions*, 2, 18

APPENDIX II

SELECTED PUBLICATIONS OF GEORGE GORDON

1 1832. The existence of blue clay on the southern side of the Murray Firth, *Trans. Geol. Soc. Lond.,* Series 2, 2, 487

2 1839. *Collectanea for a Flora of Moray,* Elgin, privately printed

3 1844. A fauna of Moray, *The Zoologist,* 2, 421–28 & 502–515

4 1852. A list of fishes that have been found in the Moray Firth and in the fresh water of the Province of Moray, *The Zoologist,* 10, 3454–62; 3480–3489

5 1852. A list of Crustaceans in the Moray Firth, *The Zoologist,* 10, 3678–87

6 1852. On the Hymenoptera (but chiefly the genus *Bombus*) of the Province of Moray, *The Scottish Naturalist*

7 1853. List of Echinodermata hitherto met with in the Moray Firth, *The Zoologist,* 11, 3781–85

8 1854. A list of Mollusca hitherto found in the Province of Moray and in the Moray Firth, *The Zoologist,* 12, 4300–4318 & 4421–4435 & 4453–4462

9 1859. On the geology of the lower and northern part of the province of Moray; its history, present state of inquiry, and points for future examination, *Edin. New Phil. Jrnl.,* New Series, 9, 14–60

10 1861. List of Lepidoptera hitherto found within the province of Moray etc.: arranged according to Doubleday's list, second edition, *The Zoologist,* 19, 7663–75

11 1863. On the relations of the Ross–Shire sandstones containing reptiliferous footprints, *PGSL,* 506–510 (with J.M. Joass).

12 1873. Introductory Note to *An Account of the Great Floods of August 1829 in the Province of Moray and Adjoining Districts* by Sir Thomas Dick Lauder (Third Edition), Elgin: J.M.'Gillivray and Son.

13 1893. The reptiliferous sandstone of Elgin, *Trans. Edin. Geol. Soc.,* 6, 241–45

APPENDIX III

SELECT BIBLIOGRAPHY

Agassiz, Elizabeth Cary. 1885. *Louis Agassiz. His Life and Correspondence.* 2 vols. Boston: Houghton Mifflin.

Agassiz, J. L. R. 1833–1843 (1844). *Recherches sur les Poissons Fossiles.* Text (5 vols.; I xlix + 188 pp., II xii + 310 + 336 pp., III viii + 390 pp., IV xvi + 296 pp., V xii + 122 + 160 pp.) and Atlas (5 vols.; I 10 pl., II 149 pl., III 83 pl., IV 61 pl., V 91 pl.) Neuchatel (aux frais de l'auteur), Imprimerie de Petit-pierre, Lithographie de H. Nicolet.

_____. 1835. "On the fossil fishes of Scotland," *BA,* 1834 Edinburgh, 646–649.

_____. 1843. "Report on the fossil fishes of the Devonian System or Old Red Sandstone," *BA,* 1842 Manchester, 80–88.

_____. 1844–1845. *Monographie des Poissons Fossiles du Vieux Grès Rouge ou Système Devonien (Old Red Sandstone) des Iles Britanniques et de Russie.* Text, xxxvi + 171 pp.; Atlas, 43 pl. 2 vols. Neuchatel (aux frais de l'auteur), Soleure, chez Jent et Gassmann.

Ahlberg, P. E. 1991. "Tetrapod or near-tetrapod fossils from the Upper Devonian of Scotland," *Nature,* 354: 298–301.

Allen, David E. 1976. *The Naturalist in Britain: A Social History,* London.

Anderson, 182?, Manuscript sketch of the Brora coal-field: read before the Philosophical Society of Inverness.

Anderson, G. 1822. "Geognostical Sketch of Part of the Great Glen of Scotland," *Memoirs of the Wernerian Natural History Society,* 4, 190.

Anderson, G. and Anderson, P. 1834. *Guide to the Highlands and Islands of Scotland.* London: John Murray.

Anderson, G. and Anderson, P. 1850. *Guide to the Highlands and Islands of Scotland.* Third Edition, Edinburgh: Adam and Charles Black.

Anderson, J. 1859. Dura Den. *A monograph of the yellow sandstone and its remarkable fossil remains.* Edinburgh: Constable.

Andrews, S. M. 1982. *The Discovery of Fossil Fishes in Scotland Up to 1845 with Checklists of Agassiz's Figured Specimens.* Edinburgh: Royal Scottish Museum Studies.

Ansted, David. 1844. *Geology Introductory, Descriptive and Practical.* 2 volumes. London: John van Voorst.

Bailey, E. 1963. *Charles Lyell.* New York: Doubleday.

Bakewell, Robert. 1828. *An Introduction to Geology Comprising the Elements of Science in its Present Advanced State, and All the Recent Discoveries; With an Outline of the Geology of England and Wales.* London: Longman, Rees, Orme, Brown and Green.

Bald. "General account of the Clackmannan and other Scottish coal-fields," *Mem. Wern. Soc. Edin.*, 3, 138.

Banks, Iain. 1992 *The Crow Road.* London: Abacus.

Barber, A. J. et al. 1978. *The Lewisian and Torridonian Rocks of North-West Scotland.* London: The Geologists Association.

Barber, Lynn. 1982. *The Heyday of Natural History 1820–1871.* London: Jonathan Cape.

Bayne, P. 1871. *The Life and Letters of Hugh Miller,* 2 volumes, London: Strachan.

Beckles, S. H. 1859. "On fossil foot-prints in the sandstone at Cumming-stone," *QJGS,* 15: 461.

Benton, M. J. 1977. *The Elgin reptiles.* Elgin: The Moray Society.

_____. 1983a. "Progressionism in the 1850s: Lyell, Owen, Mantell and the Elgin fossil reptile *Leptopleuron (Telerpeton),*" *ANH,* 11: 123–136.

_____. 1983b. "The Triassic reptile *Hyperodapedon* from Elgin: functional morphology and relationships," *Philosophical Transactions.* 302: 605–717.

_____. 1980. "The Elgin reptiles and the mid-19th-century progressionist controversy," *Mesozoic Vertebrate Life,* 1: 41–46.

_____. and Walker, A. 1985. "Palaeoecology, taphonomy, and dating of Permo-Triassic reptiles from Elgin, north-east Scotland," *Palaeontology.* 28: 207–234.

Boué, Aimé. 1820. *Essai géologique sur l'Ecosse.* Paris: Courcier.

Boulenger, G. A. 1903. "On reptilian remains from the Trias of Elgin," *Philosophical Transactions.* 196: 175–189.

_____. 1904. "On the characters and affinities of the Triassic reptile *Telerpeton elginense,*" *PZSL,* 1904, 470–480.

Brannigan, Augustine. 1981. *The Social Basis of Scientific Discoveries,* Cambridge.

Brickenden, L. B. 1850. "Fossil foot-prints of Moray," *Elgin Courant,* 18 Oct. 1850, p. 2.

_____. 1852. Notice of the discovery of reptilian foot-tracks and remains in the Old Red or Devonian strata of Moray. *QJGS,* 8: 100–105.

Buckland, William. 1823. *Reliquiae Diluviana; or, Observations on the Organic Remains Contained in Caves, Fissures, and Diluvial Gravel, and on Other Geological Phenomena, Attesting the Action of an Universal Deluge.* London: John Murray.

_____. 1836. *Geology and Mineralogy Considered With Reference to Natural Theology (Bridgewater Treatise).* 2 volumes, London: Pickering.

Burckhardt, R. 1900. "On Hyperodapedon," *Geol. Mag.* (4) 7: 486–492, 529–535.

Cadell, Henry M. 1876. *The Geology and Scenery of Scotland.* Edinburgh: David Douglas.

Calloway, C. 1883. "The age of the newer gneissic rocks of the northern Highlands," *QJGS*, 39: 355–422.

Cameron, D. 1883. "The granite and the junction of the Old Red Sandstone and the gneiss in lower Strath-nairn (Daviot)," *Trans. Edin. Geol. Soc.,* 4: 98.

_____. 1883. "The glaciers of lower Strath-nairn," *Trans. Edin. Geol. Soc.,* 4: 160.

Challinor, John. 1971. *The History of British Geology: A Bibliographical Study.* New York: Barnes and Noble.

Clemmensen, L. B. 1987. Complex star dunes and associated aeolian bedforms, Hopeman Sandstone (Permo-Triassic) Moray Firth Basin, Scotland. From Frostick, L. and Reid, I. (eds.) *Desert Sediments: Ancient and Modern,* Geological Society Special Publication No. 35, 213–231.

Coleman, William. 1964. *Georges Cuvier A Study of the History of Evolution Theory.* Cambridge: Harvard University Press.

Collie, Michael. 1991. *Huxley at Work: With the Scientific Correspondence of T. H. Huxley and the Rev. Dr. George Gordon of Birnie, near Elgin.* London: Macmillan Press.

Conybeare, W. D. and Phillips, W. 1822. *Outlines of the Geology of England and Wales.* London: Phillips.

Craig, Gordon Y. 1991. *The Geology of Scotland,* Third edition, Edinburgh: Scottish Academic Press.

Cumming, Constance. 1876. *From the Hebrides to the Himalayas,* 2 volumes.

Cunningham, Robert James Hay. 1839. "Geognosy of Scotland," *Trans. High. Agr. Soc.,* 13, 37.

_____. 1839. "On the geognosy of the isle of Eigg," *Mem. Wern. Soc. Edin.* 8: 144–163.

_____. 1841. "Geognostical account of the county of Sutherlandshire," *Trans. High. Agr. Soc.,* 7: 73–114.

_____. 1843. "Geognostical account of Banffshire," *Trans. High. Agr. Soc.* 8: 447–502.

_____. 1843. "Geognostical description of the Stewartry of Kirkcudbright," *Trans. High. Agr. Soc.* 8: 697–738.

Cutter, Eric. 1974. "Sir Archibald Geikie: A Bibliography," *JSBNH.* 7: 1–18.

_____. 1813. *Essay on the Theory of the Earth . . . with Mineralogical Notes, and an Account of Cuvier's Geological Discoveries by Professor Jameson.* First Edition translated by Robert Kerr, Edinburgh: William Blackwood.

_____. 1821. *Recherches sur les ossements fossiles.* Nouvelle edition, 6 vols. Paris: Dufour et D'Ocagne.

_____. 1827. *Essay on the Theory of the Earth, by Baron Cuvier, with Geological Illustrations by Professor Jameson.* Fifth Edition. Edinburgh: William Blackwood; London: T. Cadell.

Cuvier, G. & Brogniart, A. 1811. "Essai sur la géographie minéralogique des environs de Paris," *Mem. Classe Sci. Math. Phys., Inst. Imp. France.* 1810, pt. 1.

Davies, G. L. 1968. "The tour of the British Isles made by Louis Agassiz in 1840," *Ann. Sci.,* 24: 131–146.

Delair, J.B. & Sarjeant, W. A. S. 1985. "History and bibliography of the study of fossil vertebrate footprints in the British Isles: supplement 1973–1983," *Palaeogeography, Palaeoclimatology, Palaeoecology,* 49: 12–160.

Desmond, Adrian. 1982. *Archetypes and Ancestors. Palaeontology in Victorian London, 1850–1875.* Chicago: University of Chicago Press.

Desmond, Adrian, and Moore, J. R. 1991, *Darwin.* London: Michael Joseph.

Duff, M. E. G. 1858. "Morayshire," *Westminster Review.*

Duff, Patrick. 1842. *Sketch of the Geology of Moray, (with a map, two sections and eight plates).* Elgin: Forsyth and Young.

Egerton, P. de M. G. 1860. "Remarks on the nomenclature of the Devonian fishes," *QJGS:* 16: 119–136.

Elie de Beaumont, J. B. 1831. "Researches on some of the revolutions which have taken place on the surface of the globe," *Philosophical Magazine.* new series 10: 241–264.

Etheridge, R. 1888. *Fossils of the British Islands — Palaeozoic.* Oxford: Clarendon Press.

Eyles, Victor A. 1937. "John MacCulloch F. R. S. and his Geological Map," *Ann. Sci.* 2: 114–129.

Forbes, Edward. 1849–52. *Figures and Descriptions of British Organic Remains, Decades 1–4.* London: Geological Survey.

Frostick, L., Reid, I., Jarvis, J. and Eardley, H. 1988. "Triassic sediments of the Inner Moray Firth, Scotland: early rift deposits," *JGSL.* 145: 235–248.

Fuller, J.G.C.M. 1992. "The invention and first use of stratigraphic cross-sections by John Strachey, F.R.S., (1671–1743)," *ANH,* 19 (1): 69–90.

Geikie, Archibald. "On the Old Red Sandstone of western Europe," *Trans. Roy. Soc. Edin.* 28: 345–452.

_____. "On the phenomena of the glacial drift of Scotland," *Transactions Geological Society of Glasgow.* 1: 1–190.

_____. 1858. *The Story of a Boulder or Gleanings from the Notebook of a Field Geologist.* Edinburgh: Thomas Constable and Co.

_____. 1865. *The Scenery of Scotland*. London: Macmillan.

_____. 1875. *The Life of Roderick I. Murchison , Based on His Journals and Letters with Notices of His Scientific Contemporaries and a Sketch of the Rise and Growth of Palaeozoic Geology in Britain.* 2 volumes, London: John Murray.

_____. 1888. "Report on the recent work of the Geological Survey in the north-west Highlands of Scotland, based on the field-notes and maps of Messrs. B. N. Peach, J. Horne, W. Gunn, C. T. Clough, L. Hinxman and H. M. Cadell," *QJGS.* 44: 378–441.

_____. 1897. *The Founders of Geology*. London: Macmillan.

Geikie, J. 1874. *The Great Ice Age,* London: Isbister.

Gillen, C. "The Geology and Landscape of Moray," in *Moray Province and People.*

Glennie, K. W. (ed.) 1984. *Introduction to the Petroleum Geology of the North Sea.* Oxford: Blackwell.

Gowing, Margaret. "Science, technology and education: England 1970," *Oxford Review of Education.* 4: 3–17.

Gruber, Jacob W. and Thackray, John C. 1992. *Richard Owen Commemoration.* London: Natural History Museum Publications.

Guntau, Martin. "The emergence of geology as a scientific discipline," *History of Science.* 16: 280–290.

Harkness, R. 1859. "Of the Yellow Sandstones of Elgin and Lossiemouth," *British Association Proceedings.,* Aberdeen.

_____. 1864. "On the reptiliferous rocks and the footprint-bearing strata of the north-east of Scotland," *QJGS.* 20: 429–443.

Hedrick, Rev. James. "The Mineralogy, etc. etc of the Island of Arran," p. 221.

Helps, Arthur (ed.). 1873. *Leaves from the Journal of Our Life in the Highlands for 1848 to 1861.* London: Smith, Elder, pp. 124–6.

Hickling, G. 1909. "British Permian footprints," *Memoirs and Proceedings of the Manchester Literary and Philosophical Society.* 53: 1–31.

Hinxman, L. W. and Wilson, J. S. Grant. 1902. "The Geology of Lower Strathspey," *Memoir of the Geological Survey.*

Hooker, William. 1821. *Flora Scotica; or a Description of Scottish Plants Arranged Both According to the Artificial and Natural Methods.* In Two Parts. London: Archibald Constable and Hurst, Robinson & Co.

Horne, J. 1880. "Geology of the Nairn and Findhorn," *Trans. Inver. Sc. Soc. & Fld.* Club. 1: 283.

_____. 1893. "The Geology of Nairnside," *Trans. Inver. Sc. Soc. & Fld.* Club, 3: 51.

_____. 1894a. "The scientific work of the late Dr. Gordon, of Birnie," *The Inverness Courier.* January 12th 1894. Reprinted *Trans. Inver. Sc. Soc. & Fld. Club.* 4 (1888–1895): 294–300, with portrait.

_____. 1894b. "Obituary notice of the late Rev. George Gordon, LL.D., of Birnie," *Proc. Roy. Phys. Soc. Edin.* 12: 355–356.

_____. et al. 1923. "The geology of the lower Findhorn and lower Strath Nairn including part of the Black Isle near Fortrose," *Mem. Geol. Surv. Scotland.* 84 and part 94.

Huene, F. Von. 1908. "On the age of the reptile faunas contained in the Magnesian conglomerate at Bristol and in the Elgin Sandstone," *Geol. Mag.*, (5) 5: 99–100.

Huxley, T. H. 1859a. "Postscript [to Murchison (1859)]," *QJGS*, 15: 435–436.

_____. 1859b. "On the *Stagonolepis Robertsoni*; and on the recently discovered footmarks in the sandstones of Cummingstone," *QJGS.* 15: 440–460.

_____. 1861. "Preliminary essay upon the systematic arrangement of the fishes of the Devonian Epoch," *Mem. Geol. Surv. U.K.: Figures and Descriptions Illustrative of British Organic Remains; Scientific Memoirs.* vol. 2, pp. 4211–60.

_____. 1867. "On a new specimen of *Telerpeton elginense*," *QJGS.* 23: 77–84.

_____. 1869. "On *Hyperodapedon*," *QJGS.* 25: 138–152.

_____. 1875. "On *Stagonolepis Robertsoni*, and on the evolution of the Crocodilia," *QJGS.* 31: 423–438.

_____. 1877. "The crocodilian remains found in the Elgin Sandstones, with remarks on the ichnites of Cummingstone," *Memoir Geological Survey U.K., Monograph* 3: 1–52.

_____. 1887. "Further observations upon *Hyperodapedon gordoni*," *QJGS.* 43: 675–694.

Jameson, Robert. 1800. *Mineralogy of the Scottish Isles.* 2 volumes, Edinburgh.

Jamieson, T. F. 1865. "The history of the last glacial changes in Scotland," *QJGS.* 21: 161–203.

Jolly, W. 1878. "Excursion to Alves and Burghead," *Tran. Inver. Sc. Soc. & Fld. Club.* 1: 157–165.

Jones, Rupert. 1862. *Monograph of the Fossil Estheriae.* London: The Palaeontographical Society.

Jones, T. R. 1863. "A monograph of the fossil Estheriae," *Palaeontographical Society.* 1–134.

Judd, J. W. 1873. "The Secondary rocks of Scotland," *QJGS.* 29: 97–195.

_____. 1885. "The presence of the remains of *Dicynodon* in the Triassic sandstone of Elgin," *Nature.* 32: 573.

_____. 1886. "On the relation of the reptiliferous sandstone of Elgin to the Upper Old Red Sandstone," *Proceedings Royal Society London (B)*, 39: 394–404.

Lapworth, C. "On the close of the Highland controversy," *Geol. Mag.*, 22: 97–106.

Lauder, Sir Thomas Dick. 1830. *An Account of the Great Floods of August 1829, in the Province of Moray, and Adjoining Districts,* Edinburgh: Adam Black.

Lurie, Edward. 1960. *Louis Agassiz: A Life In Science.* Chicago: The University of Chicago Press.

Lyell, Mrs. 1881. *Life, Letters and Journals of Sir Charles Lyell.* London: John Murray.

Lyell, Charles. 1830–1833. *Principles of Geology.* London: John Murray.

_____. 1838. *Elements of Geology.* London: John Murray.

_____. 1852. *A Manual of Elementary Geology: or the Ancient Changes of the Earth and its Inhabitants, As Illustrated by Geological Monuments.* 4th Edition, London, John Murray, 512 pp.

MacCulloch, J. A. 1819. *A Description of the Western Islands of Scotland Including the Isle of Man.* London: Constable.

_____. 1821. *A Geologic Classification of Rocks With Descriptive Synopses of the Species and Varieties Comprising the Elements of Practical Geology.* London: Longman, Rees, Orme, Brown and Green.

_____. 1824. *The Highlands and Western Isles of Scotland.* 4 volumes, London: Longman, Hurst, Rees, Orme, Brown and Green.

_____. 1831. *A System of Geology With a Theory of the Earth, and an Explanation of Its Connexion with Sacred Records.* 2 volumes, London: Longman, Rees, Orme, Brown and Green.

Mackie, W. 1897. "The sands and sandstones of eastern Moray," *Trans. Edinb. Geol. Soc.* 7: 148–172.

_____. 1902. "The pebble-band of the Elgin Trias and its wind-worn pebbles," *BA* (1901): 650–651.

Malcolmson, J. G. 1838. "On the occurrence of Wealden strata at Linksfield, near Elgin; on the remains of fishes in the Old Red Sandstone of that neighbourhood; and on raised beaches along the adjacent coast," *PGSL.* 2: 667–9.

_____. 1842. "On the relations of the different parts of the Old Red Sandstone, in which organic remains have recently been discovered, in the counties of Murray, Nairn, Banff and Inverness," (Abstract) *PGSL.* 3: 141–144.

_____. 1859. "On the relations of the different parts of the Old Red Sandstone in which organic remains have recently been discovered, in the counties of Moray, Nairn, Banff, and Inverness," *QJGS.* 15: 336–352.

Mantell, G.A. 1852. "Description of the *Telerpeton elginense,* and observations on supposed fossil ova of batrachians in the Lower Devonian strata of Forfarshire," *QJGS.* 8: 100–109.

Martin, J. 1837. "On the geology of Morayshire (with map and coloured section)," *Prize Essays and Trans. High. Agr. Soc.* Series II, 5: 417–440.

Martin, J. 1856. "On the Northern drift, as it is developed on the Southern Shore of the Moray Firth," *Edin. New Phil. Jrnl.*, 4: 209–238.

Martin, J. C. 1860. (MS.) "A ramble among the fossiliferous beds of Moray," [9 pages; MS. held by Elgin Museum].

Miller, Hugh. 1857. *The Old Red Sandstone or New Walks in an Old Field.* Seventh Edition, Edinburgh: Nimmo.

_____. 1858. *The Cruise of the Betsey or A Summer Holiday in the Hebrides.* Edinburgh: Constable.

_____. 1861. *The Footsteps of the Creator; or, The Asterolepis of Stromness.* Edinburgh: Adam and Charles Black.

_____. 1889. *Rambles of a Geologist or Ten Thousand Miles Over the Fossil-ferous Deposits of Scotland.* Edinburgh: W. P. Nimmo, Hay and Mitchell.

Moore, C. 1860. "On the so-called Wealden Beds at Linksfield, and the rep-tiliferous sandstones of Elgin," *QJGS.* 16: 445–447.

Morrell, J. B. and Thackray, Arnold. 1981. *Gentlemen of Science: Early Years of the British Association for the Advancement of Science.* Oxford.

Murchison, Charles (ed.). 1868. *Palaeontological Memoirs and Notes of the Late Hugh Falconer A.M., M.D.,* London: Robert Hardwicke.

Murchison, R. I. 1845. *The Geology of Russia in Europe and the Ural Mountains.* 2 volumes, London: John Murray.

Murchison, R. I., and Nicol James, 1856, "Geologic Map of Europe," in Alexander Keith Johnston, *The Physical Atlas of Natural Phenomena.* 2nd ed., Edinburgh and London: Blackwood, pp. 13–16, pl. 4.

Newton, E. T. 1893. "On some new reptiles from the Elgin sandstones," *Philosophical Transactions.* 184: 431–503.

_____. 1894: "Reptiles from the Elgin sandstone.—Description of two new genera," *Philosophical Transactions.* 185: 573–607.

Nicol, James. 1844. *Guide to the Geology of Scotland: Containing an Account of the Character, Distribution, and more Interesting Appearances of its Rocks and Minerals With a Geological Map and Plates.* Edinburgh and London: Oliver & Boyd and Simkin Marshall and Co.

_____. 1855. "On the striated rocks and other evidences of ice-action observed in the north of Scotland," *BA*, Glasgow.

_____. 1857. "On the red sandstone and conglomerate and the super-posed quartz rocks, limestones, and gneiss of the north-west coast of Scotland," *QJGS.* 13: 17–39.

_____. 1859. "On the relation of the gneiss, red sandstone and quartzite in the north-west Highlands," *BA*, London: John Murray, 119–120.

_____. 1861. "On the structure of the north-west Highlands and the relations of the gneiss, red sandstone (Torridonian) and quartzite of Sutherland and Ross-Shire," *QJGS*. 17: 85–113.

Oldroyd, David R. 1990. *The Highlands Controversy. Constructing Geological Knowledge Through Field-Work in Nineteenth Century Britain*. Chicago: University of Chicago Press.

Orange, A.D. "The origins of the British Association for the Advancement of Science," *BJHS*. 5: 152–176.

Owen, Richard. 1860. *Palaeontology or a Systematic Summary of Extinct Animals and Their Geological Relations*. Edinburgh: Adam and Charles Black.

Page, David. 1845. *Rudiments of Geology*. Edinburgh: William and Robert Chambers.

Page, Leroy E., 1969. "Diluvialism and Its Critics in Great Britain in the Early Nineteenth Century," in Schneer, C.J . (ed.) *Toward a History of Geology*. Cambridge: The M.I.T. Press.

_____. "The rivalry between Charles Lyell and Roderick Murchison," *BJHS*. 9: 156–165.

Pander, C. H. 1860. *Uber die Saurodipterinen, Dendrodonten, Glyptolepiden und Cheirolepiden des devonischen Systems*. Buchdruckerei der Kaiserlichen Akademie der Wissenschaften, St Petersburg.

Peach, B. N. and Horne, J. "The glaciation of Caithness," *Proc. Roy. Phys. Soc. Edin.* 6: 316–352.

_____. 1879. "The glaciation of the Shetland Isles," *QJGS*, 35: 778–811.

_____. 1880. "The glaciation of the Orkney Islands," *QJGS*, 36: 648–663.

_____. 1899. *The Silurian Rocks of Scotland*. London: Geological Survey.

Peacock, J. D. 1966. "Contorted beds in the Permo-Triassic aeolian sandstones of Morayshire," *Bulletin Geological Survey Great Britain*, 24: 157–162.

Peacock, J. D., Berridge, N. G., Harris, A. L. and May, F. 1968. "The Geology of the Elgin District. Explanation of One-Inch Geological Sheet 95," *Memoirs of the Geological Survey of Scotland*, Edinburgh: Her Majesty's Stationery Office.

Phillips, J. G. 1886. "The Elgin sandstones," *BA*, (1885): 1023–1024.

Powrie, James. 1868. "On the connection of the Lower, Middle, and Upper Old Red Sandstones of Scotland," *Trans. Edin. Geol. Soc.*, 1: 115–132.

Prestwich, J. 1838. "Observations on the ichthyolites of Gamrie in Banffshire, and on the accompanying red conglomerates and sandstones," *PGSL*. 2: 187–188.

_____. 1840. "On the structure of the neighbourhood of Gamrie, Banffshire, particularly on the deposit containing ichthyolites," *Trans. Geol. Soc. Lond.*, Series 2, 5: 139–148.

Ramsay, A. C. 1864. "On the erosion of valleys and lakes: reply to Sir Roderick Murchison's anniversary address to the Geographical Society," *Philosophical Magazine*. (October 1864): 1–20.

Reeks, Margaret. 1920. *Register of the Associates and Old Students of the Royal School of Mines and History of the Royal School of Mines*. London: Royal School of Mines (Old Students') Association.

Rhind, W. 1839. *Sketches of the past and present state of Moray*. Edinburgh: printed by Andrew Shortrede.

_____. 1862. "Notice of reptilian fossils, Morayshire," *Proc. Roy. Phys. Soc. Edin.*, 2: 155–156.

Robertson, Alexander. 1846. "Notice of the occurrence of beds containing freshwater fossils in the Oolitic coalfield of Brora, Sutherlandshire," *PGSL*, 4: 173–174.

Robertson, A. 1847. "On the Wealden beds of Brora, Sutherlandshire, with remarks on the relations of the Wealden strata and Stonesfield Slate to the rest of the Jurassic System, and on the marine contemporary of the Wealden Series above the Portland Stone," *QJGS*. 3: 113–128.

Rogers, D. A. 1987. "Devonian correlations, environments and tectonics across the Great Glen Fault," PhD thesis: University of Cambridge.

Rogers, D. A., Marshall, J.E.A. & Astin, T.R. 1989. Devonian and later movements on the Great Glen Fault system, Scotland. *JGSL*. 146: 369–372.

Rosie, George. 1981. *Hugh Miller: Outrage and Order*. Edinburgh: Mainstream Publishing.

Rudwick, M. J. S. 1963. "The foundation of the Geological Society in London: its scheme for cooperative research and its struggle for independence," *BJHS*. 1: 325–355.

_____. 1974. "Roderick Impey Murchison," in Charles Gillispie (ed.) *Dictionary of Scientific Biography*, New York, 9: 582–585.

_____. 1976. "Levels of disagreement in the Sedgwick-Murchison controversy," *JGSL*. 132: 373–375.

_____. 1985. *The Great Devonian Controversy: The Shaping of Scientific Knowledge Among Gentlemanly Specialists*. Chicago and London: The University of Chicago Press.

Rupke, Nicholaas A. 1983. *The Great Chain of History: William Buckland and the English School of Geology 1814–49*. Oxford: Clarendon.

_____. 1994. *Richard Owen Victorian Naturalist*. New Haven: Yale University Press.

Sarjeant, W. A. S. 1974. "A history and bibliography of the study of fossil vertebrate footprints in the British Isles," *Palaeogeography, Palaeoclimatology, Palaeoecology*, 16: 265–378.

_____. 1980. *Geologists and the History of Geology: An International Bibliography from the Origins to 1978*. III, NY: Arno Press, 1768–1769.

Secord, J. A. 1982. "King of Siluria: Roderick Murchison and the Imperial theme in nineteenth-century British geology," *Victorian Studies,* 25: 413–442.

_____. 1986. *Controversy in Victorian Geology: The Cambrian-Silurian Dispute.* Princeton: Princeton University Press.

Shaw, Lachman. *The History of the Province of Moray.* 3 volumes, London: Hamilton, Adams and Co.

Smiles, Samuel. 1878. *Robert Dick. Baker of Thurso. Geologist and Botanist.* London: John Murray.

Stafford, Robert A. 1989. *Scientist of Empire: Sir Roderick Murchison, Scientific Exploration and Victorian Imperialism.* Cambridge: Cambridge University Press.

Strachey, J. 1719. "A curious description of the strata observed in the coalmines of Mendip in Somersetshire," *Royal Philosophical Society Transactions.* 30: 360, 968–973.

Strickland, H. E. 1840. "On some remarkable dikes of calcareous grit, at Eathie in Ross-shire," *Trans. Geol. Soc. Lond.,* Series 2, 5, 599–600.

Sweeting, George S. 1958. *The Geologists' Association 1858–1958: A History of the First Hundred Years.* Colchester: Benham.

Symonds, W. S. 1860. "On the physical relations of the reptiliferous sandstone of Elgin," *Edin. New Phil. Jrnl.,* 12: 95–101.

Taylor, W. 1894. "Note on Cutties Hillock reptiles," *Natural Science.* 4: 472.

_____. 1901. "Fossils in the Moray Firth area," *Trans. Inver. Sci. Soc. & Fld. Club.,* 6: 46–48.

_____. 1920. "A new locality for Triassic reptiles, with notes on the Trias found in the parishes of Urquhart and Lhanbryde, Morayshire," *Trans. Geol. Soc. Edin.* 11: 11–13.

Thackray, J. C. 1972. "Essential source material of Roderick Murchison," *JSBNH,* 6: 162–170.

_____. 1976. "The Murchison-Sedgwick controversy," *JGSL.* 132: 367–372.

_____. 1978. "R. I. Murchison's *Silurian System* (1839)," *JSBNH.* 7: 61–73.

_____. 1979. "R. I. Murchison's *Geology of Russia* (1845)," *JSBNH,* 8: 421–433.

_____. 1981. "R.I. Murchison's *Siluria* (1854 and later)," *ANH,* 10: 37–43.

Trail, J. W. H. 1894. "The late Rev. George Gordon, M.A., LL.D.," *ASNH,* (April 1894): 65–71.

Traill, T. S. 1841. "Notice of the fossil fishes found in the Old Red-Sandstone-formation of Orkney, particularly of an undescribed species, *Diplopterus Agassis. [sic],*" *Trans. Roy. Soc. Edin.* 15: 89–92.

Traquair, R. H. 1880. "Report on a collection of fossil fish remains from Nairnside, made by Messrs. Thomas D. Wallace and Alexander Ross, of Inverness," *Trans. Inver. Sc. Soc. & Fld. Club.* 1: 200

_____. 1886. "Preliminary note on a new fossil reptile recently discovered at New Spynie, near Elgin," *BA*, (1885): 1024–1025.

_____. 1895. "The extinct vertebrate animals of the Moray Firth area," in: Harvie-Brown, J. A. and Buckley, T. *A Vertebrate Fauna of the Moray Firth Basin*. Edinburgh, 2, 235–285.

Traquair, R. H., Powrie, J., and Lankester, E. R. 1868–1914. *The Fishes of the Old Red Sandstone of Britain*. London: Palaeontographical Society.

Verifier. 1878. *Scepticism in Geology and Reasons for It*. London: John Murray.

Walker, A. D. 1961. "Triassic reptiles from the Elgin area: *Stagonolepis, Dasygnathus* and their allies," *Philosophical Transactions*. 244: 103–204.

_____. 1964. "Triassic reptiles from the Elgin area: *Ornithosuchus* and the origin of carnosaurs," *Philosophical Transactions*. 248: 53–134.

_____. 1973. "The age of the Cuttie's Hillock Sandstone (Permo-Triassic) of the Elgin area," *Scott. Jl Geol*. 9: 177–183.

Wallace, T. D. 1880. "On the structural geology of Strathnairn, with a report on a collection of fossil fish-remains from Nairnside, made by Messrs. T. D. Wallace and A. Ross of Inverness," *Trans. Edin. Geol. Soc*. 3: 204.

_____. 1880. "The geology of the Nairn Valley," *Trans. Inver. Sc. Soc. & Fld. Club*. 1: 189.

_____. 1883. "Recent geological change in the Moray Firth," *Trans. Inver. Sc. Soc. & Fld. Club*. 2: 380.

_____. 1907. "Dr. John Grant Malcolmson and the Rev. Dr. Gordon, of Birnie, Elgin," *Elgin Courant*. June 15th, 1907. Reprinted in *Trans. Inver. Sc. Soc. & Fld. Club*, 8: 419–437.

Watson, D. M. S. 1909a. "On some reptilian remains from the Trias of Lossiemouth (Elgin)," *QJGS*. 65: 440.

_____. 1909b. "The 'Trias' of Moray," *Geol. Mag*. (5) 6: 102–107.

Watson D. M. S. and Hickling, G. 1914. "On the Triassic and Permian rocks of Moray," *Geol. Mag*. (6) 1: 399–402.

Westoll, T.S. 1937. "The Old Red Sandstone Fishes of the North of Scotland, particularly of Orkney and Shetland," *PGA*. 48: 13–45.

_____. 1979. "Devonian fish biostratigraphy," *In*: House, M. R., Scrutton, C. T. & Bassett, M. G. (eds) *The Devonian System—A Palaeontological Association International Symposium*. Special Papers in Palaeontology, 23: 341–353.

Whewell, W. 1840. *The Philosophy of the Inductive Sciences*. London: Parker, 2 Volumes.

Wignall, P. B., and Pickering, K. T. 1993. "Palaeoecology and sedimentology across a Jurassic fault scarp, NE Scotland," *JGSL*. 150: 323–340.

Williams, D. 1973. "The Sedimentology and Petrology of the New Red Sandstone of the Elgin Basin, North-east Scotland," Ph.D. thesis, University of Hull.

Wilson, H. E. 1985. *Down to Earth: One Hundred and Fifty Years of the British Geological Survey.* Edinburgh: Scottish Academic Press.

Wilson, L. G., 1970. *Sir Charles Lyell's Scientific Journals on the Species Question.* New Haven: Yale University Press.

_____. 1972. *Charles Lyell: The Years to 1841: The Revolution in Geology.* New Haven, Yale University Press, 553 pp.

Woodward, A. S. and Sherborn, C. D. 1890. *A Catalogue of British Fossil Vertebrata.* London: Dulau.

Young, Robert. 1871. *The Parish of Spynie in the County of Elgin.* Elgin: Printed at the Courant Office by James Black.

APPENDIX IV

OBITUARY OF RODERICK I. MURCHISON BY ARCHIBALD GEIKIE IN THE PROCEEDINGS OF THE ROYAL SOCIETY OF EDINBURGH SESSION 1871–1872

Among our recent losses there is none which we have more reason to deplore than his. The name of Sir Roderick Murchison has been a household word in geology for nearly half a century, not in Britain only, but also over all the world. While we share in the wide regret at the injury which the general cause of science sustained by his removal, we add also the sadness which arises from the recollection of the relation which he bore to the progress of geology in Scotland, and from what he has recently done for the advancement of its study in the University of this city.

Born in 1792 at Tarradale, in Ross-shire, he was educated for the military profession, and served during part of the Peninsular War. But on the arrival of peace in 1815, finding that the army no longer opened up the same prospect of activity for which he longed, he gave up his commission, married, and settled in England. The succeeding part of his life, prior to 1824, he used to speak of as his "Foxhunting period," when he threw himself with all the ardour of his nature into the field sports of a country residence. Part of that period, however, he spent abroad, making with his wife, tours in search of picture galleries and old art, and keeping an elaborate diary, with criticisms on the character of the fine arts in each tour of collection visited. It was by a kind of happy accident that his energies were at last directed into the channel of science,—the merit of which change was due partly to his wife's taste for natural history, and partly to the friendly counsel of Sir Humphry Davy. He joined the Geological Society of London, and soon became one of its most enthusiastic members. From that time forward his love for geology, and his activity in its pursuit, never waned. He travelled over every part of Britain, and year after year he resorted to the Continent, traversing it in detail from the Alps to Scandinavia, and from the coasts of France to the far bounds of the Ural Mountains. As the result of

232

these journeys, there came from his pen more than a hundred memoirs, besides two separate and classical works on 'The Silurian System,' and on 'Russia.'

Sir Roderick was essentially a geologist, and he chose one special branch as his own domain. Perhaps no man ever had the same power, —which seemed sometimes almost an intuition,—of seizing the dominant features of the geographical and palaeontological details of a district. With a keen eye to detect the characters as they rose before him, and a faculty of rapidly appreciating their significance, he could, as it were, read off the geology of a country after a few traverses only, when most men would have been puzzling over their first section. This was the secret of his broad generalisations regarding the geological structure of a large part of Europe,—generalisations which, though of course requiring to be corrected and modified by subsequent more detailed investigations, still remain true in the main, and still astound by their marvellous grasp and suggestiveness. The leading idea of his scientific life was to establish the order of succession among rocks, and through that order to show the successive stages in the history of life on our globe. With the more speculative parts of geology he meddled little; nor did he ever travel outside the bounds of his own science. He early recognised the limits within which his powers could find the fullest and most free development, and he was seldom found making even a short excursion beyond them.

The special part of his work on which his chief title to fame rests is undoubtedly his establishment of 'The Silurian System.' Before his time, the early chapters of the history of life on our globe had been but dimly deciphered. William Smith had thrown a new flood of light upon that history by showing the order of succession among the secondary rocks of England, and had done more than any other man to dispel the prejudices with which the doctrines of Werner seemed naturally to fill the mind. But the rocks older than secondary, to which Werner had given the name of 'Transition,' remained still in deep Wernerian darkness. Sir Roderick Murchison saw that it might be possible to bring order and light out of these rocks, even as had been done with those of more recent origin; and that a double interest would attach to them if, as he supposed, they should reveal to us the first beginnings of life upon our globe. Choosing a part of the broken land of England where the rocks are well exposed, he set himself to unravel their order of succession. Patiently year after year he laboured at his self-appointed task, communicating his results sometimes in writing to his friends, sometimes in the form of a short paper to the Geological Society of London, until at last, in 1839, he gathered up the whole into his great work, 'The Silurian System.' In that book the early chapters of the history of life on the earth were first unfolded, and a system of classification was chosen with such skill that it has been found applicable, with minor modifications, even in the most distant quarters of the globe.

Round this early work all his after-labours seemed to range themselves by a natural sequence. His choice had led him into the most ancient fossiliferous rocks, and to that first love he remained true. Whether in the glades of Shropshire, or the glens of his own Highlands, among the fjelds and fjords of Norway, or in the wilds of the Urals, it was with the Palaeozoic formations that he mainly busied himself. They were to him a kind of patrimony which had claims on his constant supervision. With his friend Sedgwick he unravelled the structure of the middle Palaeozoic rocks of Devonshire, and with Keyserling and De Verneuil he showed the true relations of the upper Palaeozoic rocks of Russia. The Silurian, Devonian, and Permian systems, representing each a vast cycle in the history of our earth as a habitable globe, received in this way from him their first clear elucidation, and the very names by which they are now universally known.

But if we seek to measure the influence which Sir Roderick Murchison exercised on the progress of the science of the time merely by the original work which he himself accomplished, we should fail duly to appreciate the measure and the power of that influence, and the extent of the loss which his death has caused. Fortunate in the possession of wealth and high social position, he was enabled to act as a constant friend and guardian to the cause of science. He moved about as one of the representative scientific men of his day. To no man more than to him do we owe the public recognition of the claims of scientific culture in this country. For he not only stood out as the acknowledged chief in his own domain, but had also the faculty of gathering round him men of all sciences, among whom his kindliness of nature, his courteous dignity of manners, his tact and knowledge of the world, and his wide range of social connections marked him out as spokesman and leader. Nowhere were these features of his character and influence more conspicuous than in his conduct of the affairs of the Geographical Society, of which he was for many years the very life and soul, and which owes in large measure to him the stimulus it has given to geographical science.

Here in his own native country, and more especially here in Edinburgh, we have peculiar cause to mourn the loss of such a man.

Though his residence from boyhood had been chiefly in London, he never to the last relinquished his enthusiastic regard for the land of his birth. He never lost an opportunity of boasting that he was a Scot. During the last ten years of his life he made frequent and protracted tours in the Highlands; and, in unraveling their complicated geological structure, he accomplished one of the most brilliant generalisations of this long and illustrious scientific career. There is something touching in the reflection that, after having travelled and toiled all over Europe, gaining the highest position and rewards which a scientific man can attain, he should at last, ripe in years and in honours, have come back to his own Highlands, and there completed his life-work by bringing into order the chaos of the primary rocks, and laying such an impress on Scottish geology as had never been laid before by any single observer. For these and other researches he received from this Society the first Brisbane Medal—an honour conferred on him at the Aberdeen meeting of the British Association, and of which he often spoke as one that gave him the deepest gratification. He used to boast, too, of being an honorary Fellow of this Society, and to quote a remark made to him by the late Robert Brown, that his election into the list of our honorary Fellows was one of the highest marks of distinction he could receive. His kindly interest in our prosperity was often expressed; and we have a token of it in the presentation to us of his bust by Weeks, which this evening is formally delivered to the Society.

Of the closing acts of his life, there is one which cannot be mentioned without peculiar pride—the institution of a Chair of Geology and Mineralogy in the University of Edinburgh. He intended to found this Chair by bequest; but on the retirement of Dr Allman from the Chair of Natural History, he determined to do in his lifetime what would otherwise have been accomplished not till after his death. He gave to the University a sum of £6000; and the Crown having consented to add an annual grant of £200, the Chair was founded in the spring of the present year. Sir Roderick has not lived to witness the first beginnings of the tuition which he had started. But long after the memory of his personal character shall fade, men will remember the work which he did; they will recognise the impetus his researches have given to geology all over the world; and let us hope also they will see in the Chair he has founded the starting-point of a new and active school of Scottish geology.

APPENDIX V

ON THE GEOLOGICAL STRUCTURE AND ORDER OF THE OLDER ROCKS IN THE NORTHERNMOST COUNTIES OF SCOTLAND

Sir R. Murchison delivered a lecture on the above subject to a large audience, in the Music Hall, on Friday last—Sir D. Brewster presiding.

In a brief introduction, Sir Roderick, alluding to his early military career, apologized for having undertaken, at the request of his friends, to describe the order and succession of the various older rocks of his native Highlands, and expressed the hope that the same indulgence would be shown towards him by his countrymen on this occasion, as was manifested at the Manchester meeting seventeen years ago, when he endeavoured to convey, in a popular manner, his view of the geological structure of the Empire of Russia. He then went on to explain the progress which had been made in the classification of the rocks of sedimentary origin in Scotland. First, alluding to the great leaders of Scottish geology—Hutton, Playfair, and Hall, and his immediate predecessors—Jameson, M'Culloch, and others—he showed to how great an extent the chief point on which he was going to insist—the metamorphism of sedimentary strata of various ages into crystalline rocks had been ably illustrated by Hutton himself. After his day, however, mineralogy chiefly occupied the minds of geologists, and comparatively little progress was made for some years in geology as at present cultivated. With William Smith, however, a new era arose in England, and the proofs which that sagacious man brought forward to show that each sedimentary formation was characterised by organic remains peculiar to it, and that there existed a regular order of superposition from the older to the younger strata were the true foundations or keystones of modern geology. So long as this invaluable and truthful doctrine was applied to the secondary and tertiary formations of England, which are, for the most part, slightly disturbed, little difficulty occurred in comparing portions of any distant set of rocks with their types in a well-known and well-searched district. Thus, thirty-three years ago, he (Sir Roderick, beginning with a Scottish illustration) having served his apprenticeship under his eminent leader, William Smith, in the examination of the cliffs of the Yorkshire coast, where certain seams of impure coal crop out from amidst strata replete with fossils of the oolitic series[1], had no difficulty in assigning to the long stripe of land on the East coast of Sutherland the same age as the sandstones, shale, and coal of the East of Yorkshire, and also in proving that sediments of like age also occupied large portions of the Hebrides. In this way an accurate approximation was commenced to be made between some English and Scottish sediments of secondary age. In the following year, Professor Sedgwick and himself, devoting a long summer to the North of Scotland, endeavoured to prove, that the Old Red Sandstone of Scotland was simply the equivalent of that of England, and they pointed out the relations and component parts of that great group in Caithness, Ross, Moray, &c. They then also made an observation of considerable importance, and which has had a direct bearing on the main object of the present discourse. They then saw that the quartz rocks and limestones of Sutherland were overlaid by a great series of micuous[2], quartzose, and talcose schists and flags, all of which, together with some varieties of flaggy gneiss, were ranged in that primary and crystalline series in which no organic remains had then been detected. For a long time this observation lay in oblivion; since it could have no valuable application until a discovery should be made of organic remains in some of the bands of rock. In truth, Professor Sedgwick and himself had not then entertained the idea of eliminating the order, succession, and fossil contents of the oldest fossiliferous rocks of England and Wales. But when their classification of such rocks was established and found to be true over many and wide regions of the world, including Russia, it was also found to apply to the Southern Counties of Scotland, which, from their bold character, have been called the South Scottish Highlands. In the extension of the Silurian classification to the South of Scotland, it gave him pleasure to state that the first realization of the view proceeded from Professor Nicol, who is so serviceable at this meeting, and who is justly esteemed as the Professor of Natural History in Marischal College of Aberdeen. Having suggested that the day would come when many of the crystallized schists, sandstones, and limestones of the North Highlands, would also prove to be simply the altered

---

[1] In a number of places in this account, the reporter failed to capitalize stratigraphic terms such as geologic periods and systems. The account as transcribed includes those errors.

[2] A mispelled version of micaceous.

or metamorphosed representative of the silurian rocks of the South of Scotland. Sir Roderick felt the strongest anticipation that by some good chance or other organic remains would be discovered in those rocks, and this desire was not long kept in abeyance. For, in the following year, his friend Mr C. Peach, who, when doing duty on the extreme coast of Cornwall, had previously discovered Lower Silurian fossils in hard quartzites, had the good fortune, whilst visiting a wrecked ship in Durness, to discover traces of organic life in the limestones of that tract. On seeing these fragments, Sir Roderick, fully aware of their stratigraphical position, pronounced them to be Silurian forms. As, however, he could not then induce many persons to adopt that view, he requested Professor Nicol to accompany him to the North-West of Sutherland in 1856, and in the following autumn (1856) he communicated his own opinion thereon, and on the whole succession of the crystalline strata to the Meeting of the British Association at Glasgow, when he presided over the Geological Section. On that occasion, his friend Professor Nicol expressed doubts, which he has since abandoned, as to the fossil-bearing strata being of Silurian age, and subsequently, having revisited the tracts, and greatly extended his sphere of observation, he published his views in the Journal of the Geological Society, and expressed the opinion that the fossiliferous limestones of the West of Sutherland were probably of a much younger age and might belong to the carboniferous period. Being opposed to such an inference, Sir Roderick naturally requested his friend Mr Peach to lose no opportunity of completing the fossiliferous proofs on which, and on the order of superposition his opinion was based, and the result was the discovery of many more, and much more clearly defined fossils than had hitherto been found, and thus no doubt of their being of Lower Silurian age could be entertained. Sir Roderick then pointed to an enlarged diagram in which these various fossil shells were represented—explained their places in the Natural History Kingdom, and how several of them were species also known in the Calciferous Sand rock, and Bird's Eye limestone group of the Lower Silurian rocks of North America, and eulogised the palaeontological skill of Mr Salter, who had figured, described, and identified them at his request. He then stated that during the last year he revisited all these tracts for the third time, and that as he was then accompanied by Mr Peach, he propounded his matured views to the Geological Society in 1858, and the results were before the public. In determining the relative ages of the rocks in a country like the North of Scotland, where most of the masses have been subjected to great changes from their original condition, and have been much metamorphosed, the first requisite is to obtain any one clear and determinate horizon, the relative place of which in the general series has been ascertained in other parts of the world. Now, as this has been accomplished by the registration of the age of these Durness fossils, and that they proved the deposits to be very low in the Silurian series, so it followed, that any great mass of sediment which lay beneath the strata containing such remains, must belong to a pre-existing period. Referring to a large diagram showing the position of the underlying rocks, and also to a pictorial view of the mountains as seen from the sea near Loch Inver, he explained how the Lower Silurian quartz rocks, with their subordinate limestones, reposed on great mountain masses of red, purple, and chocolate-coloured sandstone, grit, and conglomerate. These masses of vast thickness were referred by Sir Roderick to the Cambrian age; and he showed that they must have undergone powerful movements and great denudation before the Lower Silurian fossiliferous quartz rocks and limestones were deposited upon them. On the point of unconformable or transgressive superposition, as shown in a pictorial diagram, he expressed his obligation to Colonel James and Professor Nicol, who, in the summer of 1856, when he was not in Scotland, had extended his views, and clearly defined that feature of inconformability, as respected the mountain of Queenaig and others, in the west coast of Sutherland. In terming these sandstones and grits Cambrian, he affirmed that, even mineralogically, they were identical in aspect with those of like age in the Longmynd of Shropshire, and of Harlech in North Wales, whilst the quartz rocks which over-laid them were undistinguishable from the quartz rocks of the Stiper Stones, which, twenty-five years ago, he fixed upon as the base of the Silurian rocks, both the English and the Scottish rocks being also characterised by having been perforated by annelids or sea-worms. As his countrymen were naturally proud of the antiquity of many things belonging to their native land, he was glad to tell them that the most north-western portion of their Highlands contained rocks of higher antiquity than any which had been discovered in England, Wales, or Ireland. In all those other parts of the British Isles, the most venerable rock which could be detected (and the countries had been well explored) were the Cambrian rocks, which, as he showed (pointing to one of the numerous diagrams), are second in the order of Scottish Rocks. The fundamental gneiss of the north-west of Scotland is, he believed, also older than any rock in France, Germany, Italy, or Spain. He showed, on the authority of Professor Ramsay, who had recently explored North America, that this fundamental British gneiss was the probable

equivalent of the great Laurentian System, so defined in Canada by
Sir William Logan. He then dwelt upon the mineral constitution
and aspect of this old or fundamental gneiss which lay at the base
of the North British rocks, and which he unequivocally separated
from another and more flag-like gneiss, which was seen to occur far
above the fossiliferous Lower Silurian rocks. It was chiefly to assure
himself that his old section of thirty-two years back, and confirmed
by two subsequent visits, that he had this year made a fourth survey of these rocks of the north-western Highlands; and, in order to
have the case tested by a geologist as honest as he is skilful in field
work, he induced Professor Ramsay to accompany him. So plain
and irresistible were the evidences presented in a number of trans-
versed sections at Loch Broom, Loch Assynt, Loch More, Loch Eribol, &c., that the only wonder entertained by his companion was
that any scepticism should have prevailed as to the order of succession. In illustration of this ascending order from the Lower Silurian
group of quartz rocks, with their included limestones, he specially
instanced the cases which occurred at the upper end of Loch Assynt,
where the limestone, fairly encased in quartz rocks, both beneath
and above it, was splendidly exhibited in Ben More of Assynt. There
the upper zone of quartz rock, descending from these great heights
into the lower country, watered by the river Oykel, is surmounted
by another band of limestone, which, in turn, is clearly and conformably surmounted by the micacious[3] talcose, chloritic, and other
schistose rocks seen in descending the Oykel river from Sutherland
into Ross-shire. After giving illustrations of a similar succession in
Glencoul and Loch More, Sir Roderick particularly dwelt on the
clearness of the order as exposed on the east bank of Loch Eribol;
and here, if any one should doubt that the limestones and their including quartz rocks were not of the same age as those of Durness,
he held up in his hand a specimen of an orthoceratite from Eribol,
presented by his friend Mr Clark, and which was actually imbedded in the quartz rock of that place. Now, in pursuing a traverse of
the mountains which separate Loch Eribol from Loch Tongue, clear
proofs were seen that the quartz rocks and limestones, with their
fossils, were regularly and symmetrically covered by micacious
flags, in parts talcose, in parts approaching to a sort of gneiss. This
order is not only observed by passing along the northern road
which traverses the moors called the Maen, but is still more clearly
exhibited by taking the road which, striking across the head of Loch
Hope, leads to Altnaharrow. In fact, the whole mountain of Ben

------

[3]A variation on the term micaceaus.

Hope (3150 feet above the sea) consists of these crystalline mica-cious flagstones—all of which, for a space of many miles, are per-fectly comformable in dip and strike to the purer quartzose group with fossils into which they graduate downwards; the whole series being more or less silicious. Having reason to think that the locali-ties where certain local igneous rocks (felspathic, quartzose, granitic, and porphyritic) had been intruded among the otherwise regular strata, might be dwelt upon as indications of a great break which would weaken the proofs of succession. The author and Pro-fessor Ramsay paid special attention to this point, and everywhere found that such intrusions, though productive of local disturbances and partial alterations, caused no derangement of the order of suc-cession beyond the limited areas in which such agency had been rife. The most striking example of these intrusions of felspathic rocks is seen in the promontory of the Whiton Head made well known to tourists by Walter Scott, and which they examined minutely; and even there, notwithstanding powerful intrusions of felspathic granite, the deep green chloritic and micacious schists were observed to repose conformably upon those white quartz rocks, from which the headland takes its name, and which are in truth merely the northern extensions of the rocks of Eribol in which the orthoceratite was found. The same local intrusion and interfer-ence is again seen at one spot between the Whiten Head and Eribol, and several others are traceable along this line from N.NE. to S.SW., in one place bursting through the quartz rock, in another through the overlying micacious schist; but nowhere, including the flanks of Ben More of Assynt, do such intrusions constitute a geological sep-aration between two groups of rocks. For, such igneously formed rocks, it was stated, were to be detected throughout nearly the whole ascending series. Thus, the earliest of them is the large crys-tallized porphyry of Canisp which Mr Peach first observed, and which occurs between the fundamental gneiss and the Cambrian sandstone; others are associated with the Silurian quartz rock and limestone; others again appear on the line between the fossiliferous Silurians and those overlaying strata also of Silurian age, which in the guise of mica schists, talcose and quartzose flags, and even as a sort of gneiss, have afforded no fossils. In proceeding from the north-west to the south-east, the prevalent dip being always in the latter direction, these flagstones usually become more and more gneissose; and this change in the superior portions of the masses is considered by the authors to be referable to the profusion of gran-ite eruptions, which have so diversified the eastern portions of the North of Scotland, and have so broken up the stratified masses as to render any attempt at a definition of their order most difficult, particularly in Aberdeenshire. But the Aberdenians have had their full recompense in the possession of splendid quarries of granite (some of which Nicol believes to be of high antiquity), which had not only enabled them to rear the magnificent city in which they were assembled, but which supplied London itself with materials for so many useful as well an ornamental purposes of life. The geologist who, like himself, had to delve out the order of the earlier periods of nature was compelled, therefore, to get away as far as possible from those tracts, whether in the *eastern* districts of Sutherland, Ross, Inverness, and Aberdeenshire, and where the old historical legends were so much metamorphosed that infinite labour would be required to disentangle the relative ages of the stratified rocks. Unwilling, for the present, to pass from the order exhibited in the succession of the older rocks of the western parts of Sutherland, Ross, and the adjacent parts of Inverness-shire, where their history is so clear, he could not, however, forbear from speculating on the probability that the great zone of quartz rocks and limestones which range from Jura and Islay on the SW., to Aberdeenshire and Banff-shire on the NE., may prove to be repetitions of the Lower Silurian rocks of the NW. of Sutherland. He thought it also a fair surmise that the micacious flagstone of the Duke of Richmond's quarries near Fochabers and the clay slates of [Foudeland] were both members of the Silurian series. In alluding to the distinctions which Dr M'Culloch had ably drawn between the different varieties of crystallized rocks, and eulogizing his successful efforts in developing the igneous origin of granites, and the transition of trap rocks into true volcanic substances, he warmly commended the labours of the late Mr Cunningham in his survey of Sutherland. That gentleman clearly advocated much of the doctrine as proved by natural evidences, on which Sir Roderick now insisted; but the map of that gentleman, like the last geological maps of Scotland, invariably represented all gneiss (whatever was its age), under one colour, whilst the sandstones of the west and those of the east coast, now known to be of such very different ages, were also represented under one tint of old red sandstone. In the second part of the discourse, the old red sandstone, properly so called, was described as existing on the eastern flanks only of the stratified crystalline rocks, and as being made up of their debris. This old red group consists of three zones, the lowest being conglomerates and sandstones, the middle the Caithness flags, and the upper sandstones, for the most part yellow. In treating of this part of the subject, Sir Roderick warmly eulogized the labours of his lamented friend, Hugh Miller, and expressed his satisfaction at having seen, in a recent excursion to Cromarty, the monument erected to the memory of that truly eminent man. He regretted, indeed, that Agassiz, the great naturalist, who had described so many of the fossil fishes of this age, and who had come from his adopted country (America) to be once more among them at a meeting of the British Association, had been suddenly called away by domestic affliction. Maintaining that the Scottish old red was the full equivalent of all the strata termed "Devonian" by Sedgwick and himself, he showed that, in the northern Highland counties, the lower zone had afforded no fossils, but that in Forfarshire. In alluding to the Lower Old Red, or Zone characterised by the presence of the *Cephalapsis*, Sir Roderick praised highly the researches of the Rev. Mr Mitchell, who had produced at this meeting several forms of fishes and vegetables from the deposit in Forfarshire, and had read a clear account of them. The species are all new, according to Sir Egerton, but one of the genera is the *Acauthodes*[4] *of Agassiz*. It had yielded the same fossils as the lowest old red of Herefordshire and Shropshire, where it lies upon, and passes down into, the uppermost Silurian rock. In speaking of the numerous great fractures which had affected the rocks of the north of Scotland, which could not be dwelt upon at a popular meeting and in a brief discourse, he pointed out the fact that the whole of the coast of Eastern Ross, or the Black Isle of Ross-shire, from the Sutors of Cromarty to Avoch, had been affected by an intruded band of igneous rock, by which not only the older schists had been much metamorphosed, and had assumed the character of gneiss, but by which the conglomorates[5] and sandstones forming the base of the old red group had been thrown up into vertical and mural forms. He expressed his belief that this great fracture from N.NE. to S.SW was seen to extend across Scotland, and was marked by the great depression occupied by lochs, which the Caledonian canal has united. The Caithness flags, constituting the central fossiliferous zone, had been so well illustrated by Hugh Miller, and this part of the subject was now so well known, that he then called attention to the chief features of the group as is ranged along the northern shores of Nairn and parts of Banffshire. In that range the lower conglomorates were well seen in the deep southern recesses of the streams which flow from the adjacent mountains, and he cited fine examples of this rock to be seen at and above Cawdor Castle and other places. The middle zone is slenderly exhibited, if certain argillaceous bands with nodules filled

---

[4]A misspelling of *Acanthodes.*
[5]A misspelling of the term conglomerates.

with ichthyolites as at Lethen Bar, Clune, Tynet, Gamrie, &c., were alone to be viewed as their equivalents, seeing that they are in truth charged with many of the same species of *coccostens*[6], *osteolepis*, *pteraspis, asterolepis, cherolepis*[7], and *cheiracanthus*, &c., which abound in Caithness. But in truth some of these fossils, particularly the pteraspis, range upwards of Nairn and Moray into beds where they are associated with numerous forms of *Holoptychius,* a genus scarcely known in Caithness and the Orkney Islands. This inter-mixture is of value in showing that in Moray the type of the old red sandstone of Perthshire and Fifeshire, where *Holoptychii* abound, was already visible as we proceed from the north to the south. The chief interest, attached to these beds charged with *Holoptychii,* as seen to the north of the town of Elgin, consisted in their passing directly under and appearing to be intimately connected, as far as the drift-covered surface permits of examination, with other sand-stones in which the reptiles *Stagonolepis Robertsoni,* the *Hyperodape-don Gordoni* of Huxley, and the *Telerpeton Elginense* of Mantell had been detected. When it was first announced that, according to the stratigraphical evidences of these reptilian strata seemed to form an upper part of the old red sandstone, a great sensation was naturally produced among palaeontologists and geologists. Though Profes-sor Sedgwick and Sir Roderick had thirty two years ago included all the yellow sandstones of Elgin in the old red deposit, the discovery of the *Telerpeton* in that rock some years ago astonished geologists, who were surprised that an air-breathing lizard should have been found in a rock of such high antiquity. Still more were palaeon-tologists astonished when after revisiting the Elgin tract last year, Sir Roderick announced the discovery of bones of another reptile whose scutes, which had formerly been supposed to be the scales of fishes, were with singular ability proved by Professor Huxley to be-long to a group of reptiles allied to crocodiles. Again, this surprise was intensely increased when the Rev. George Gordon discovered in this sandstone the remains of another reptile closely resembling the Rhyncosaurus of the Trias, and considered by Huxley to be of the lacertian division of reptiles. In consequence of the scepticism which prevailed concerning the true geological habitat of these rep-tiles, Sir Roderick revisited the spots accompanied by Professor Ramsay, immediately before the meeting, and came away under the conviction that no stratigraphical evidence exists by which the strata containing the reptiles can be separated from those charged

---

[6]A misspelling of *coccosteus.*
[7]A misspelling of *cheirolepis.*

with *Holoptychii* and other fossil fishes. Whilst it is right to state that
no absolute junction between the fish beds and the reptile beds has
been detected in the ridge of Elgin (that tract being very much covered by drift, and the rocks being exposed in quarries only), still
as the yellow and white sandstone with fish scales of Bishopmill,
Elgin dips gently to the N.NW., and is succeeded in perfect conformity at a very short distance by the reptile beds which overlie it at
Findrassie, also dipping at the same angle to the N.NW., any field
geologist would say that they belonged to the same natural physical group. Both sets of beds are chiefly composed of yellow and
white sandstones, which colours, with a few exceptions of red and
green tints, as seen at the Findhorn and other places, are the prevalent colours of the so-called old red sandstone throughout the Elgin
tract. Again, both sets of beds, whether red, yellow, or white, are associated with cornstone[8]; the lower of *Holoptychian* strata overlying
the cornstone to the south of Elgin, these at Cothall, on the Findhorn, and many other places, being largely burnt for lime, whilst the
reptilian sandstone, which everywhere has the same strike and dip
as the subjacent fish beds, is in its turn overlaid by another cornstone with a very silicious matrix. Although it could not be posi-
tively demonstrated that these reptilian strata should be classed
with the old red sandstone since they have not yet yielded any fossil characteristic of that age, he suggested that from the conformity
and similarity of the two, the uppermost which had alone offered
reptiles of a crocodilian type might lead us to suppose, as thrown
out by his associate, Professor Ramsay, that the seas in which the
Holoptychii swam, had been succeeded by a fresh water, conditions
in which the marine fishes could no longer live, though the one deposit was simply the continuation of the other. If this should prove
to be the case, the transition (Sir R. added) was similar to that which
occurred in the secondary period, when the marine oolite deposits
were succeeded by the lacustrine and fluviatile animals of Purbeck;
or, as when in tertiary times the sea bottoms, replete with shells of
miocene age, as in the south of Russia, of a fresh water, or brackish
water fauna; or again, as in the present day in South America, where
lagoons inhabited by alligators, are in juxtaposition to the ocean.
But no such theory would, he was well aware, satisfy the objections
of his friends, those palaeontologists who from all former experience, were led to believe in a progression of creation from lower to
higher types, and who, until an absolute demonstration was given,

---

[8]This term was printed repeatedly as corustone in the newspaper article. Perhaps the reporter had difficulty in distinguishing between Murchison's n and his u.

could not be brought to think, that crocodilian and lacertian reptiles, which, as yet, are unknown in the Cortoniferous[9] and Permian deposits of younger age than the old red sandstone, can really have existed in that remote period. It was, however, remarkable that none of the genera of reptiles found in the Elgin sandstones had ever been discovered in any secondary rock. Whilst the reptilian sandstones occupy only the north-eastern and depressed portion of the Elgin ridge, the same band is much more extensively developed in the coast ridge which extends from Burghead to Lossiemouth, where a succession of splendid quarries of the finest building stones are opened out. In fact, the town of Elgin, with its beautiful old Cathedral, elegant villas, and numerous public buildings, exhibits an instructive juxtaposition of these admirable freestones, which, from their sectile qualities when first extracted, and their hardening quality under the atmosphere, their imperviousness to wet, and their great durability, are strongly to be recommended for use in the metropolis of London, where their warm tints, varying from rose-colour to yellow and white, would be greatly admired. In examining the various public buildings of Elgin, some of which are built of the fish-bearing stone and other of the reptiliferous stone, any lithologist would be puzzled to draw a distinction between them. He (Sir Roderick) was bound to state the results of his two surveys of the Elgin district, though he would much rather from analogy and prepossession in favour of a progressive order of creation, have been enabled to offer to his friends the palaeontologists some stratigraphical reason why these reptilian strata might be of the age of the trias or new red sandstone. There was, indeed, another reason why the reptilian sandstone could not well be supposed to belong to the trias. On various parts of the low country of Moray, patches of secondary strata with numerous lias fossils had been detected. Some of these had been referred to lias, others to inferior, and Oxford oolite, like the deposits on the east coast of Sutherland; some of them containing impure coal like that of Brora. Now, it is quite evident that these deposits (now almost entirely denuded), must have extended transgressively over all the older rocks of the region, for some of them are found lying on members of the true old red sandstone with its fishes, others on the reptiliferous strata; Mr Gordon and Professor Ramsay having detected a patch on the shore near Lossiemouth, let in by a fault, and abutting abruptly against the cornstones of the reptile beds. If, then, the reptilian sandstones were of the age of the

---

[9]Carboniferous is meant. Presumably the reporter had difficulty reading Murchison's scrawl.

Trias, it is very improbable, if not impossible, that whilst the true old red is conformable to such sandstones, the lias which ought to succeed them in the regular geological order is entirely transgressive to them. Sir Roderick then duly complimented the valuable services rendered to geological science by the gentlemen of Elgin, more particularly by Mr P. Duff and the Rev. G. Gordon of Birnie, the latter having been his guide and instructor on three several occasions, and who is a very good observer himself, is unhesitatingly of opinion, as well as Mr Duff, that the reptiliferous and fish bearing sandstones are physically united. Lastly, as regarded the Old Red Sandstone, he spoke of a considerable Aberdeenshire development of the formation in Strathbogie, to the south of Rhynie, where Professor Ramsay and himself, led thither by their friend, the Rev. Alex. Mackay, had observed such a succession of strata as gave them reason to think that in one part of the deposit fossil fishes, like those of Fochabers and Tynet and Gamrie, might be discovered; whilst in the superior sandstones, often of white colours, Mr. Mackay had discovered casts of fossil stems of large land plants, rudely resembling calamites. It was ascertained (Sir R. observed) that land plants of considerable size, of which there were no traces in the Silurian epoch, had begun to grow during the Old Red or Devonian times; and he pointed to a pictorial diagram representing a great insular ocean, which might be considered to be Scotland in a very ancient period, when large plants grew upon its shores, and reptiles crawled on sand and mud banks, or nestled among those early groves. This imaginary view might with more propriety be applied to Scotland when after the completion of the Old Red deposits, the Grampians and adjacent mountains formed an island, on the southern shores of which that arboreal vegetation prevailed, consisting of an abundance of vascular plants, referable to treeferns, palms, calamites, and others out of which the coal of the Fife, Edinburgh, and Glasgow coal-fields had been elaborated. Unluckily for his native Highlands, they seem to have been excluded, by some great physical operation, from those conditions, for they offer no trace of the carboniferous or coal-bearing deposits. These, and those deposits which succeeded to them, and to which he had applied the term "Permian" (from an immense region of Russia over which they spread), have never, it would appear, been formed to the north of the Grampians. Of the secondary deposits, he had already briefly alluded to the existence of the lias and oolitic formations, which, slenderly exhibited only on the east coast, were very largely developed in the Hebrides, patches of them having been discovered in Aberdeenshire. If time had permitted, he might have said a few words on these accumulations, as well as on those remains of chalk-flints described by Mr Jamieson, of Ellon, and others. Again, he might have occupied an evening in lecturing on the great superficial masses of detritus and huge erratic blocks which were distributed over Scotland during the glacial period. But those subjects had been amply illustrated by other authors, and his main object was to direct attention exclusively to the foundation-stones of Scotland and their true order. In offering this sketch of some of the leading features of what he called a "Reform Bill" of the geological structure of the northern counties of Scotland, Sir Roderick called the attention of the Assembly to the progress which was being made by the geological survey of Great Britain under his direction, and under the special management in the field of his friend, Professor Ramsay. Exhibiting certain sheets of maps, on the six-inch scale, of the counties of Edinburgh, Haddington, and Linlithgow, which explained the outcrop of the coal and limestone of these tracts, he trusted that the staff of geological surveyors at present allotted to Scotland would be soon augmented; and, in that case, he hoped to live to see the day (if maps were only provided) when all the geology of Aberdeenshire and the North of Scotland might really be worked out with accuracy. This present effort was chiefly confided to the application of recognized general principles of classification, to the elucidation of the order of the older rocks of the Highlands and nothing more could be attempted until a country possessed maps— the North of Scotland being almost the only country in Europe without an accurate map (a melancholy fact on which he insisted a quarter of a century ago, when the Association met in Edinburgh in 1834). On that occasion the Association, at his request, memorialized the then Government; and this state of matters is being rapidly wiped away as regards all the tracts to the south of the Grampians, and he hoped that the skill and energy of his friend, Colonel James, and the officers under him, would be so warmly supported by Parliament and the public that Scotland would have before long a really good topographical map, without which no practically useful geological results can be worked out. Sir Roderick concluded his address by impressing upon the minds of his auditors who were not geologists the nature of the great difference between the formerly accepted notion of the order and equivalents of the older rocks of the North of Scotland, and those which he desired to establish by his reform, by pointing to two generalized diagrams. One of these, representing the old notions, exhibited a great central mass of rocks, termed gneiss, mica schist, quartz rocks, with granites, porphyries, &c. flanked both on the east and the west coasts by old red conglomerates and sandstones. The other, on which he had previously lectured, exhibited the succession which had been ascertained to prevail, and the order which had been evolved out of that which was previously an assemblage of crystalline rocks, distinguished only by their mineral characters, but undefined by their relative position and imbedded organic remains, and in which the rocks of the north-west coast were confused with those of the east coast.

In the course of the lecture, Sir Roderick spoke in a tone of generous appreciation of the labours of Professors Sedgewick and Nicol, Mr Peach, &c., &c. As illustrations he had a generalised section across the North of Scotland, and other diagrams. The lecture was most attentively listened to by a very full audience.

At the close—

Sir Charles Lyell moved a vote of thanks to Sir Roderick for the clear and admirable illustration he had given them of the Geology of Scotland. Without giving an opinion as to the Elgin sandstones, he (Sir C.) might state that, if the evidence so clearly laid before them bearing thereon had been before him when he published his work, he certainly should not have called that rock Devonian. Professor Phillips seconded, remarking on the high estimation in which Sir R. Murchison was held over half the globe as the master of the Silurian.

Sir D. Brewster then, amid loud applause, thanked Sir R. Murchison, and continuing, said a more lasting testimonial than a mere vote of thanks was now to be presented—viz., the Brisbane Medal of the Edinburgh Royal Society (of which Sir David is a Vice-President.) A deputation of members of the Council were now in waiting for this purpose, whom he begged to introduce. The deputation consisted of Dr Christison, Professors Allman and Balfour, and Mr R. Chambers. The medal, amid renewed plaudits, was presented through Sir David Brewster.

APPENDIX VI
SECTION C—GEOLOGY
MONDAY, SEPTEMBER 19, 1859

The feature of the geological section was Professor Nicol's paper on the Gneiss, Red Sandstone and Quartzite of the North-west Highlands and the discussion to which it led. The learned Professor stated that he had spent several summers investigating the order of the super position of the rocks in that region, and therefore he hoped his conclusions would be thought entitled to some consideration. He was sorry to appear in opposition to his friend Sir Roderick Murchison, but as he (Sir R. M.) had thrown down the gauntlet last year at Leeds, he was bound to stand up in defence of his opinions that these rocks were of igneous origin. Sir Roderick replied, his position which he defended at considerable length being simply this—that he believed these rocks Silurian. Professor Sedgwick spoke, inclining towards Professor Nicol's view, while Professor Ramsay of the Museum of Practical Geology, London, defended the opinion of Sir Roderick Murchison. The result was that the matter was left for farther consideration.

Professor Huxley, Museum of Practical Geology, London, next read a paper on the newly discovered reptilian remains from the neighbourhood of Elgin. He commenced by describing their discovery, and paid a very high compliment to the Rev. Mr. Gordon of Birnie, but for whom, he said, they should not have yet had these valuable fossils. One of these fossils, the *Stagonolepis*, was considered by Agassiz, when first shown to him, to be a fish, but further investigation had clearly proved to him (Professor Huxley,) that it was a reptile of the crocodilian type. He showed how he came to this conclusion: —First, by a comparison of the fossil scutes with the scutes of living crocodiles; secondly, by the tail; thirdly, by the jaws; and fourthly, by the vertebrae. The other reptile was the *Hyperodapedon*, specifically named *Gordoni*, in honour of the minister of Birnie. The address was illustrated by the specimens on which his conclusions were founded, and large drawings of different parts of the animals.

There was little time left for discussion on this or any of the re-maining papers, the subject of which were as follows: —'The Yellow Sandstone of Elgin and Lossiemouth,' by Professor Harkness; 'The Age of the Reptile Sandstone of Morayshire,' by J. Miller, Esq.; 'The Supposed Wealden and Other Beds Near Elgin,' by C Moore, Esq.; 'Dura Den Sandstone,' by the Rev. Dr Anderson, Newburgh.

APPENDIX VII

INVERNESS COURIER CORRESPONDENCE

To: *Inverness Courier*
From:
Symonds
Dated:
16 December 1862

I am one of those 'field geologists' who examined the physical geology and the district, in company with the Rev. G. Gordon and several other geologists, after the meeting of the British Association in Aberdeen, and as I hold *very strong doubts* respecting the assumed Devonian, or Old Red, age of the reptiliferous beds of Spynie, Lossiemouth, &c., perhaps you will do me the favour of inserting the following communication.

All thanks are due to the Rev. G. Campbell of Tarbat, and the Rev. Mr. Joass of Edderton, for their perseverance and good work in having detected the *reptiliferous deposits* at Portmahomack and other localities on the north-west coast of the Moray Firth, although I do not think that any physical geologist, who knows the Elgin district, ever doubted those rocks would in time be detected there. When I visited my friend, Mrs. Hugh Miller, at Cromarty, after our investigation of the Elgin district, I particularly requested her to try and induce her geological friends and acquaintances to search for the reptiliferous deposits in the district around Cromarty, Tain, and other localities. It would give me, Sir, real gratification if the *Telerpeton, Stagonolepis,* and *Hyperodapedon Gordoni* could be proved satisfactorily to be reptiles which were contemporaneous with the deposition of the far-famed old red sandstone of Hugh Miller, or even of the lower carboniferous deposits; but I am compelled to declare, although I cannot enter into particulars in a communication of this kind, that I have no faith in the truth of such a correlation, and I do not think that the discovery of reptiliferous deposits on the north-west side of the Moray Firth has advanced this correlation by a single step. It is my belief that when the very difficult geology of the district in question is worked out by the geological surveyors, or exposed by railways, it will prove to be somewhat similar to the maintops in the neighbourhood of Weston-super-Mare, and other localities of the S.W. of England, where rocks of the age of carboniferous limestone are seen in conjunction with those of the age of the upper trias and lias, and where fossils of palaeozoic and mesozoic character may be obtained almost in contact with each other, as in the railway cutting at

Uphill, near Weston. In the Elgin district also, it is my opinion, that the reptiliferous deposits, whether Permian or Triassic, are of much later date
than the Devonian or old red deposits that they overlie. As regards the
stratigraphical reading of the rocks from Portmahomack, by Geanies, to
Shandwick and Cromarty, permit me to state, that the only reading which
will throw light upon the question, will be the interpretation of those hitherto hidden pages, the *actual points of contact* between the reptiliferous deposits and the underlying old red rocks. The similarity in mineralogical
character and colour, between the *Holoptychian* sandstones and the reptiliferous deposits, will not allow of any other reading in such a difficult district. My brother geologists in the North of Scotland will then, I trust,
forgive me if I say that I do not consider that they have as yet proved the
Devonian age of the reptiliferous deposits. They have, I believe, great difficulties to contend with, and I am quite prepared to hear that *Holoptychian*
scales have been found in the *same quarry* with reptilian footprints or reptilian scales or bones, but they must bring forward a rock specimen containing, in its matrix, reptilian relics embedded with the remains of Old
Red fish; and they must show us a section exhibiting a conformable, up-
ward, continuation from the *Holoptychian* sandstones into the reptiliferous
rocks, before we can accept their present stratigraphical reading of the Elginshire and Ross-shire deposits. When they have done this, no geologist
will rejoice more with them than your obedient servant.

Geology of [Ross Shire]

To the Editor of the Inverness Courier.

Sir—Although the time for fighting out the battle of Tarbat has not yet
arrived, you will kindly allow me to fire off a pistol in acknowledgement
of the long—[shot] of the Rev. Wm. Symonds. We humbly submit that in
his demands for evidence of the Devonian age of the Tarbat Ichnites he is
too exacting both as to kind and degree. The proof from conformable sequence we hope to be able to give, but to produce a cabinet specimen or
any single slab "containing in its matrix reptilian relics embedded with the
remains of old red fish" would, under the circumstances, amount to evidence more precise than that upon which most of the existing classifica-
tions of geologists depends, more conclusive perhaps than that on which
those rocks have been grouped into a separate system, which may yet have
to be restored to the Upper Silurian and Lower Carboniferous.

It is by no means our intention to consider the question in its present form
as settled in favour of the palaeozoic age of the Ichnites, but we shall, with
all deference, continue to regard it as open, until careful examination of a
cliff-section, nowhere masked from top to bottom, enable us to connect the
Geanies Ichthyolites with the Ichnites of Portmahomack, when we shall

probably feel justified in occupying higher ground. Meantime should scutes or bones or *holoptychian* scales be found among the sun-cracked slabs of the Ichnite quarry, the [true] pioneers who are now at work will expect to be believed, if they assert that they have found them [there] as their sole object is to collect and preserve as much as possible in situ, such evidence as may be worthy the attention of more advanced geologists who may care [to] visit the place.

I am, Sir, yours obediently,

A Pioneer [the Rev. James Joass]

APPENDIX VIII
CORRESPONDENCE

Year	Day	Month	Sent to	Reference
Year	*Day*	*Month*	*Sent to*	*Reference*
1839	4	January	RIM to Malcolmson	DDW 71/839/1
1858	5	September	to Gordon	858/20
	14	September	to Murchison	21
	8	October	to Gordon	23
	10	October	to Gordon	24
	12	October	to Gordon	25
	14	October	to Murchison	26
	20	October	to Murchison	27
	25	October	to Murchison	28
	29	October	to Gordon	30
	2	November	to Murchison	31
	5	November	to Gordon	32
	9	November	to Murchison	33
	21	November	to Gordon	37
	25	November	to Gordon	38
	27	November	to Gordon	40
	30	November	to Murchison	41
	1	December	to Gordon	42
	5	December	to Murchison	LGS/M/G10/1
	5	December	to Gordon	43
	17	December	to Gordon	46
		December	to Gordon	47
	21	December	Gordon to Balfour	48
	23	December	to Murchison	48
	30	December	to Gordon	858/50
1859	1	January	to Murchison	859/1
	12	January	John Anderson to Gordon	3
	10	February	to Gordon	7
	14	February	to Murchison	LGS/M/G10/2
	14	February	to Murchison	8
	7	March	to Murchison	11
	11	March	to Gordon	12
	15	March	to Murchison	13

Year	Day	Month	Sent to	Reference
	12	April	to Murchison	18
	14	April	to Gordon	19
	18	April	to Murchison	21
	[18	May]	to Murchison	58
	12	June	to Gordon	30
	16	June	to Murchison	31
	21	July	Murchison to Lyell	M/L17/29
	25	August	to Gordon	28
		August	Gordon to Falconer	7
	6	October	to Gordon	35
		October	to Murchison	42
	12	October	Murchison to Lyell	ALS: BD25.L
	19	October	Murchison to Lyell	ALS: BD25.L
	29	October	Robert McAndrew	41
	22	November	to Murchison	LGS/M/G10/3
	22	November	to Murchison	47
	1	December	to Gordon	52
1861	14	May	to Gordon	861/3
	27	May	to Murchison	3
1862	10	June	to Gordon	862/12
		July	to Murchison	12
	30	July	to Murchison	12
	2	August	to Murchison	17
	8	November	Joass to Stables	38
	21	November	to Murchison	40
	24	November	to Gordon	41A
	15	December	Joass to Murchison	M/J5/1
	19	December	to Gordon	49
	21	December	to Gordon	51
	31	December	Gordon to Symonds	Lyell Papers: U. of Edin. 5646
1863	2	January	Symonds to Lyell	U. of Edin. 5645
	17	January	Joass to Gordon	863/6
	21	January	Joass to Gordon	7
	16	April	to Murchison	38
	19	April	to Gordon	39
	21	April	to Murchison	LGS/M/G10/4
	24	April	to Gordon	42
	15	May	to Murchison	50
	20	May	to Gordon	54
	18	June	to Gordon	62
	13	July	to Gordon	863[ ]
	16	July	to Murchison	—

Year	Day	Month	Sent to	Reference
1865	29	December	to Murchison	LGS/M/G10/5
1866	24	July	to Murchison	867/26
	30	July	to Gordon	866/25
	13	September	to Gordon	37
1867	24	January	to Gordon (2 copies)	867/3
		January	to Murchison	866/8
	31	January	to Gordon	8
	31	January	James Nicol to Gordon	7
	7	November	to Gordon	25
	8	November	to Murchison	869/25
	27	November	to Murchison	LGS/M/G10/6
	2	December	Nicol to Gordon	867/30
	5	December	Nicol to Gordon	30(b)

INDEX